THE Math with a Laugh SERIES

Venn Can We Be Friends?

& Other Skill-Building Math Activities

GRADES 6–7

Faye Nisonoff Ruopp & Paula Poundstone

HEINEMANN
Portsmouth, NH

Heinemann
A division of Reed Elsevier Inc.
361 Hanover Street
Portsmouth, NH 03801–3912
www.heinemann.com

Offices and agents throughout the world

Library of Congress Cataloging-in-Publication Data
Ruopp, Faye Nisonoff.
Venn can we be friends? : and other skill-building math activities, grades 6–7 / Faye Nisonoff Ruopp & Paula Poundstone.
p. cm. — (The math with a laugh series)
Includes bibliographical references.
ISBN 0-325-00927-9
1. Mathematics—Study and teaching (Middle School)—Activity programs.
I. Poundstone, Paula. II. Title. III. Series: Ruopp, Faye Nisonoff. Math with a Laugh series.
QA135.6 .R864
510.71'2—dc22 2006010850

Editor: Leigh Peake
Production: Abigail M. Heim
Typesetter: Gina Poirier Design
Cover and interior design: Joni Doherty Design
Cover and interior illustrations: Michael Kline (www.dogfoose.com)
Manufacturing: Louise Richardson

Printed in the United States of America on acid-free paper
10 09 08 07 06 VP 1 2 3 4 5

To Charlie & Marcus,
for making my life infinitely joyful.

—Faye

To Toshia, Allcy & Thomas E,
without whom nothing adds up. Thank you.

—Paula

Contents

Patterns, Relations & Algebra

Geometry

Measurement

Data Analysis, Statistics & Probability

Preface

When we first decided to collaborate on a mathematics book, we had in mind the creation of problems to be done during the summer. Schools have a long tradition of assigning summer reading; many teachers ask for parallel assignments in mathematics so that students do not lose ground over the summer months. Doing math in the summer—what a thought! Of course, many students will wonder why anyone would create math problems for vacation time. Believe it or not, we're sympathetic to that feeling. And that's why we've created a set of problems that we hope will be different from those found in standard textbooks—different in tone and style, but not in content. These problems are intentionally silly and humorous so students can laugh and be serious about the mathematics, all at the same time!

There has actually been some analysis of the benefits of humor in mathematics classrooms. In the December 2004/January 2005 edition of *Mathematics Teaching in the Middle School*, George and Janette Gadanidis and Alyssa Huang state,

> There are several benefits to using humor in the mathematics classroom (Cornett 1986, 2001; Dyer 1997; Martin and Baksh 1995; Medgyes 2002; Wischnewski 1986):
>
> • Humor helps create a more positive learning environment. It helps reduce barriers to communication and increase rapport between teacher and students.
>
> • Humor helps gain students' attention and keep their interest in a classroom activity.
>
> • By reducing stress and anxiety, humor helps improve comprehension and cognitive retention.
>
> • Humor improves students' attitudes toward the subject.
>
> • Humor helps communicate to students that it is okay for them to be creative; to take chances; to look at things in an offbeat way; and perhaps, even make mistakes in the process.

• Humor can help students see concepts in a new light and increase their understanding.

• The use of humor is rewarding for the teacher, knowing that students are listening with enjoyment. (10 [5]:245)

Although designed for use in the summer, these problems can also serve as a supplement to the curriculum during the academic year, as math to do at home with parents, as well as for skills reinforcement. Students need a change of pace and environment at times. These problems were created to provide entertaining contexts while keeping the mathematics content targeted and sound. The problems can be used as assessments, assignments, additional practice, or extra credit, as well as summer work. In addition, you will note as you scan the problems that there is a good deal of reading involved, making them an excellent tool for students to practice reading in context. We assume, then, that these materials could also be used for reading practice with students.

We ended up with a series of three books, one each for grades 4 and 5, 6 and 7, and 8 and 9. The content for these grade-level books is based on the focus areas identified in state and national standards. These areas, however, may vary from school to school. You may therefore choose to use problems from different grade-level books to accommodate your needs. Our goal was to make the materials as flexible as possible.

Whenever we look at mathematics materials, we tend to be curious about the authors, wanting to know who they are and why they wrote the materials at hand. So we've each included a short piece about ourselves, since we think our story is one that may both surprise and entertain you.

From Faye Nisonoff Ruopp

Paula Poundstone was a student of mine in the 1970s at Lincoln-Sudbury Regional High School in Sudbury, Massachusetts. Paula would say that she was never very good at math; I would say quite the contrary. I saw potential. Paula went on to be a highly successful comedian after she graduated, and we have remained close over the past thirty-two years. Paula now has her own children who are studying mathematics, and at times, I get calls (some of them late at night) about how to do some of the math problems they get in school. Once Paula told me that she made up stories for the problems to make them easier for her children to understand. Given her comedic talents, these stories turned out to be gems. And that's when the idea of collaborating on these books occurred to us. So now, after thirty-two years, she and I can proudly say that she has written a math book with her math teacher, an accomplishment that makes us both smile. We've come full circle, and we think this book is symbolic, in many ways, of the special relationships that students and teachers form, of the humanity that characterizes the study of mathematics, and of the belief that all students can learn and enjoy doing mathematics—and even smile through it all!

Many teachers hope to make mathematics playful and friendly for their students. I would like to extend the opportunity to parents as well. In thinking about my experiences as a parent doing mathematics together with my son, Marcus (who has far surpassed my mathematical abilities, I am proud to admit), I recall fondly the times when we sat down together to tackle a tough problem and the car rides when I posed problems such as "We've decided you can go to bed a half hour later each year. At some point you won't be going to sleep at all. How old will you be then?" He worked on that problem for an hour on our way to Vermont one weekend, not knowing anything about fractions. I also recall when he was about five, I asked him, "What would happen if you subtracted six from two?" His response: "You would get four in minus land!" His connection of mathematics to some fantasy world of negative numbers reminds me how important it is for children to experience their own inventions and perceptions of how mathematics makes sense to them. Likewise, Paula's fantasy contexts, rooted in humor and humanity, enable us to laugh while at the same time thinking hard about how the mathematics works.

From Paula Poundstone

How come math makes people cry? You'd think, of all subjects, history would be the tear jerker. But I cried over math when I was a kid. My mother used to cry when I asked her to help me. My high school math teacher and coauthor of this book, Faye Ruopp, kept a box of tissues on her desk, and if she ran out, class had to be canceled.

I can remember, when I was a kid, I'd get a word problem, something like: "Mary had four apples. She shared two of them with Joe. How many does she have left?"

Although I could calculate the remaining apples, I mostly wanted to know more about Mary and Joe and would often include that curiosity in my homework. Were they just friends? How did Mary get the apples? Why couldn't Joe take care of himself? What is it with Joe? Was that even his real name?

So when my own daughters were so frustrated and intimidated by their elementary school math assignments that they, too, followed the time-honored tradition of shedding buckets of tears over the wonderful world of math, I began to write personalized practice problems for them. Not surprisingly, once the problems seemed less serious, they relaxed a bit and much of the drama slipped away. We have also spent the last few summers doing a page or two of math each day and, no duh, both girls took a huge leap in their math ability as a result. I think the main thing is that it increased their confidence so they hit the ground running in the fall. We've saved lots of money on tissues and I'm hoping you will too.

I think the idea of our writing a book of these kinds of problems came from Ms. Ruopp. She had called me because she was going over her grade book from 1976 and noticed I still had some assignments missing. We got talking and I told her about doing math with my kids and the next thing you know...

And so we offer you these problems in the spirit of improving understanding and increasing rapport with your audience. We hope that when your students do these problems, they will smile and perhaps even laugh, and come to realize that mathematics can be fun and challenging and enlightening, all at once!

Acknowledgments

From Faye

My first memories of mathematics come from my paternal grandfather, Morris Nisonoff, who was a butcher in Jamesburg, New Jersey. He could add a column of numbers faster than anyone I know. I found that fascinating. My gratitude goes to him, then, for making calculations seem fun and accessible. To my own father, I express my love and gratitude for spending time doing math problems with me as a young child many mornings before I went to school. He thought a great way to start the day was to tackle two-digit multiplication! As an accountant, he too had a knack for working with numbers that transferred to both me and my sister, as we each eventually became mathematics teachers. My father, mother, and grandfather taught me the importance of doing mathematics at home with children, and the key role parents play in creating a positive disposition toward math.To that end, doing math with my own son, Marcus, has been a highlight of my parenting. I thank him, especially for continuing a tradition of math study as an applied mathematics major at Yale. His positive and joyful approach to mathematics mirrors his approach to life—how he makes me smile!

I would also like to acknowledge my past and present students, who taught me what it means to come to understand mathematics, and what it means to struggle with a subject that for many is formidable. Their spirit, humanity, diligence, and enthusiasm are continually inspiring. Teaching them has been a gift.

To my friends and family and colleagues in education who encouraged me to write this series, I thank you for your support and faith in this project. You will see yourselves in some of the problems we've created, and we hope they make you laugh.

I would like to thank Ellen Lubell for her impeccable legal expertise and advice in addressing the contractual issues, and for her support as a friend and confidante.

I extend my deepest gratitude to Leigh Peake at Heinemann, who had the vision and courage to support the initial idea for this project. I am indebted to her for her continued influence on the series. A special thanks to Michael Kline for his artistic genius in creating the cartoon illustrations, capturing the essence of the problems and adding to the spirit of the contexts. I also want to thank Abby Heim and Beth Tripp for their care, expertise, and mathematical acuity in editing the series.

And of course, my heartfelt thanks goes to my coauthor, Paula Poundstone, whose comic genius continues to inspire me. Beyond her creativity and sense of humor, she is a remarkable human being and a fabulous mother. Collaborating with Paula on this project has been infinitely rewarding—we laughed so much more than we thought we would! She has proven herself to be the mathematician I always knew she was.

And finally, I want to thank my husband, Charlie, for his unconditional support and calming influence throughout this project. As Paula's high school biology teacher, he also appreciated her amazing talent and encouraged our collaboration. This project never would have happened without him.

From Paula

The fact that I have been a part of the creation of a math book defies the laws of probability. Simple mathematical reasoning tells us that there must have been some other important factors that made this improbability possible.

I'd like to thank Leigh Peake at Heinemann for her kind support and skill. Someday I hope to remove a thorn from her paw.

I thank Abby M. Heim for making my part make sense.

I greatly appreciate the technical support of my assistant, Carmen Cannon, and that of my friend, Gordon McKee.

I will always be in the debt of my manager, Bonnie Burns, for clearing the path for me for thirteen years.

Faye Nisonoff Ruopp has been my friend, teacher, and mentor for thirty-three years. My admiration and appreciation of her grows exponentially each day. Without Faye, who knows what n would equal?

Number Sense & Operations

The problems that follow are in the Number Sense and Operations strand. The mathematics in these problems focuses on developing students' number sense as well as their computational skills. Students in grades 6 and 7 extend their knowledge of whole number operations to understanding and computing with rational numbers (fractions, decimals, and percents). They also study the order of operations and exponents.

The topics for these problems were chosen from state and national standards:

- Compare and order integers, fractions, decimals, and percents
- Apply the rules of positive integer exponents
- Understand prime, composite, prime factorization, greatest common factor (GCF), least common multiple (LCM), and divisibility rules
- Perform operations (addition, subtraction, multiplication, and division) on fractions and decimals
- Apply the order of operations
- Estimate square roots

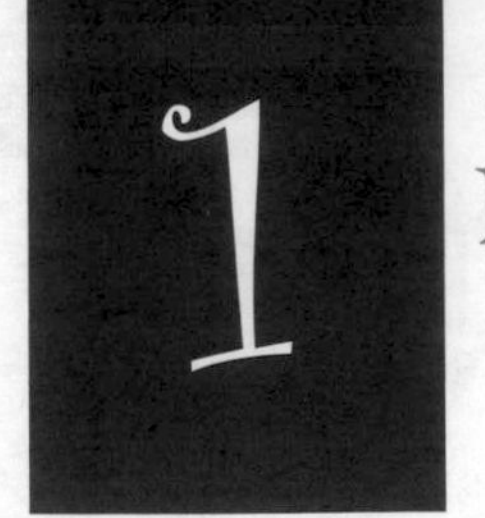

Number Sense & Operations

Problem

A Few Good Factorers

Captain Frank Factoring is the toughest drill instructor in the Division Division of the army. When he orders a soldier in training to "get down and give me fifty," that soldier drops to the ground, whips out paper and a pencil, and divides fifty problems. O'Malley is a particularly nervous new recruit. The Division Division is tougher than he thought it was going to be, and it seems like Captain Factoring has it in for him.

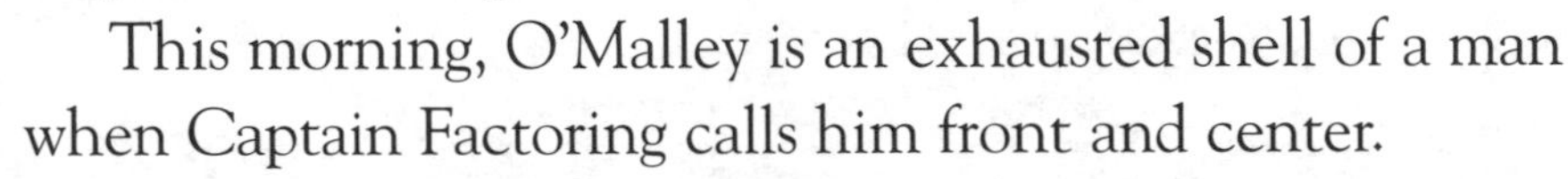

This morning, O'Malley is an exhausted shell of a man when Captain Factoring calls him front and center.

"O'Malley," the captain barks, "take the number 100."

"Yes, sir," answers O'Malley, scribbling the number frantically onto the page.

"Fac-t-o-o-or it," booms the captain.

"Sir, yes, sir." O'Malley dashes $100 = 25 \times 4$ onto his page.

"And fac-t-o-o-or those numbers!"

"Sir, yes, sir!" Trembling, O'Malley scratches $25 = 5 \times 5$ and $4 = 2 \times 2$.

"And fac-t-o-o-r those numbers again, O'Malley!"

O'Malley isn't sure he can take it anymore, but he is pretty sure the numbers can't be factored any further.

"I said, fac-t-o-o-r again, O'Malley!"

"I can't factor it anymore, sir!" stammers O'Malley, falling into a heap.

"Factor it again or you'll be factoring in the factoring factory into the next century," bellows the captain, with a little bead of sweat running straight down the center of his forehead.

"But, sir," breathes O'Malley from under his paper, "I've factored it all of the way to its prime factorization."

"Oh, you have, have you?" screams Captain Frank Factoring, with his nose practically touching O'Malley's. "Then factor 56, private!"

O'Malley is about to faint. Could you give him the prime factorization of 56 before he cracks?

Problem

Lovers and Other Numbers

On a cold, moon-bright night just after a snowfall, 9 and 12 sat on a bench at the edge of the park.

"I love snow," said 9, shyly leaning toward 12.

"Me, too," said 12 excitedly, while gently tilting his tens column toward 9.

"I like it when it gets so deep that, when I stand, it comes up to the bottom of my round part and I can disguise myself as a small zero," 9 giggled, "and I like it when it's cold."

"Me, too! We have so much in common," said 12, staring at 9 almost without breathing.

"Really?" whispered 9. "Like what else?"

"Um . . . ," stammered 12. "Well, there's bubble wrap! We both like to pop bubble wrap."

"I don't. I hate to waste resources. I worry about about the earth," said 9.

"Oh, well, then there's . . . Um, um, um," said 12, wondering how the snow could stay frozen in this heat.

Quick, name the greatest common factor and least common multiple of 9 and 12 before their special moment is lost.

The Unparalleled Thrill of Prime Numbers

Leonardo Digit hates Monday nights because his dad, Dick Digit, owner of Dick Digit's Number Depot ("Where the whole family can shop for numbers and operations don't hurt a bit!"), insists on watching *Prime Time*, which Leonardo thinks is the most boring show on television.

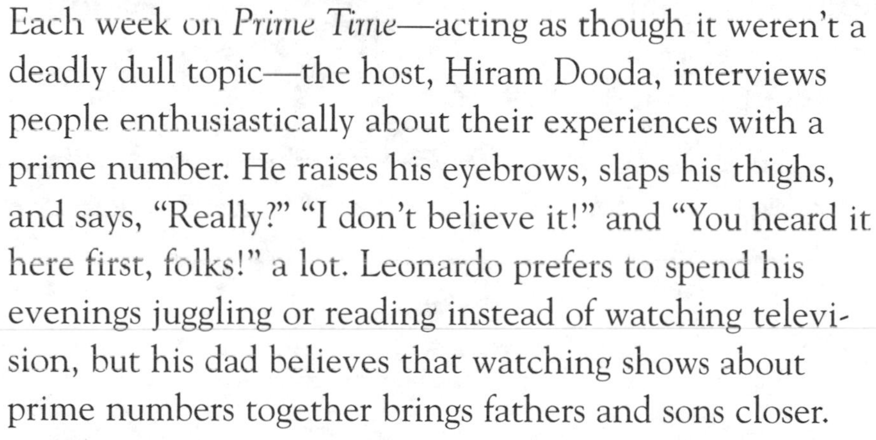

Each week on *Prime Time*—acting as though it weren't a deadly dull topic—the host, Hiram Dooda, interviews people enthusiastically about their experiences with a prime number. He raises his eyebrows, slaps his thighs, and says, "Really?" "I don't believe it!" and "You heard it here first, folks!" a lot. Leonardo prefers to spend his evenings juggling or reading instead of watching television, but his dad believes that watching shows about prime numbers together brings fathers and sons closer.

This week's memorable show was about the number 19. If Leonardo can fill his dad in on the next prime number after 19 before next week's show, maybe he can get his father to juggle with him instead of watching *Prime Time*. Leonardo is desperate. Next week's show is a two-hour special. Please help him out. What is the next prime number? How do you know?

4 Number Sense & Operations

Problem

Divide or Be Devoured

Oh, this is just great. You and your friends are being charged by nine hungry, jealous Tyrannosaurs that don't share well, and all you have to offer them is 2,123,456 Doggie Yums. If there's not enough for them to get exactly the same amount, they'll eat your little curly-headed friend who knows all of the knock-knock jokes. You start doling out the Doggie Yums, but your friend could use a head start, just in case. Quick, use the Divisibility by 9 Rule to see if 2,123,456 is divisible by 9 or if your friend has answered his last "Who's there?"

Put Out the Fire-Bellied Toads

Thomas E is six years old. Not all of the time, but more than occasionally, he forgets to think before he does stuff. He pulled out his two front teeth a long time before the tooth fairy was ready, and ideas like gravity seem to keep surprising him.

He gets a $1.50 allowance each week. Recently he was visiting a pet store and began doing karate kicks. He kicked the cover off a tank of fire-bellied toads that cost $7.50 each. The tank was 30 cm tall and the cover was 103.7 cm from the ceiling. The ceiling was 210 cm from the floor and there was a loose budgie perched on a beam up there.

When the lid clattered off, three orange and two green fire-bellied toads jumped out of the tank from the top of their fake waterfall. One customer screamed frantically, although I don't know why. They're just *called* fire-bellied toads—they're not actually a fire hazard. Thomas E captured a green one behind the foot of a screaming customer who was sitting with her head between her knees, looking pretty green herself. Perhaps the toad hopped to her for camouflage. I caught two orange ones and the third caught sight of the python and headed back to the tank on its own. The last green fire-bellied toad headed down the street toward the bagel shop and hasn't been heard from since.

Well, of course we have a situation here, and when you say to your parents and teachers, "Why do we have to do math?" let this be a lesson to you. There's a lot to sort out. Let's get going. No whining.

A. How tall in centimeters was the table the tank sat on?

B. How high was Thomas E's karate kick?

C. Of course, Thomas E has to pay for the one missing fire-bellied toad. How many weeks of allowance will it take?

D. What fraction of toads did not escape? What is this fraction as a decimal?

E. How many times did Thomas E's mother tell him not to do karate kicks?

A Brief Tale of a Long Tail

Cal is a dog with an unusually long tail. Don't say anything to him about it. He's very sensitive. He sometimes carries it in his mouth, hoping people will think it's a leash and that he's such a responsible dog that he walks himself. Jack, the neighborhood cat, made the mistake of mocking Cal about his tail. (Jack won't make *that* mistake again.)

At birth, Cal's tail was a remarkable 15.573 cm. His brother's was 15.68 cm, but their eyes weren't open yet so they didn't know. Cal's tail kept shooting out like Silly String as the rest of his body grew at a normal rate. At six weeks his tail was 31.372 cm. People used to round the measure off to the nearest hundredth when they spoke of it to make him feel better.

A. Whose tail was longer at birth, Cal's or his brother's? How much longer?

B. What was Cal's tail length at six weeks, rounded to the nearest hundredth?

C. Round Cal's tail length at six weeks to the nearest tenth. Does that make Cal feel much better? Why?

D. If Cal's brother's tail grew a remarkable 5.3 cm per week, how long was it 4.5 weeks after he was born?

Problem

Flying Cake

I love Wilbur Wright. He's Orville Wright's brother. Together they built and flew the first successful airplane. The first flight covered 120 feet in 12 seconds. That was apparently too far to walk back then. I celebrate Wilbur Wright's birthday whenever I feel like it because I don't know when it is, and he's been dead for a long time, so giving him a surprise party is out of the question.

I shared my last Wilbur Wright birthday cake with friends. Faye ate $\frac{3}{7}$ of the cake (and got the pieces with "day, Wil" and a wing made out of icing) and Charlie ate $\frac{5}{7}$ of the cake (and got "Happy Birth" and a wing and a tail made out of icing). Is this even possible?

Choose one of the following:

A. No, because I would never share that much cake.

B. No, because you'd have to use a protractor to cut a cake in sevenths, and that would wreck the icing.

C. No, because $\frac{7}{7}$ is a whole, so you'd need more than one cake to serve $\frac{3}{7}$ and $\frac{5}{7}$ of a cake.

D. Yes, you think so, but fractions make you very uncomfortable.

Daredevil Dresses

The first flying machine the Wright brothers tested out at Kitty Hawk was a glider, which is a plane without a motor. After five weeks of experimenting with their glider, they gave it one last flight. It crashed in the sand and they left it and went home to Ohio. Bill Tate, who lived in Kitty Hawk and had assisted the Wright brothers, carted the plane back to his house, where his wife, Addie, clipped the fabric from around the wooden wing frames and sewed dresses for her two daughters. The dresses were lovely but the Tate sisters often got stuck in the trees on the way to school.

If Irene, wearing the dress made from the left wing, was blown $\frac{5}{6}$ of a mile, and Pauline, wearing the dress made from the right wing, was blown $\frac{3}{8}$ of a mile, how much farther did Irene get blown?

Number Sense & Operations

Problem

Addition Prescription

Add $\frac{5}{12} + \frac{3}{5}$. Now how do you feel?

The World's Biggest Mix-Up

The best stirrers from seven countries made up the nine top competitors at this year's International Stirring Competition. They each stirred a powder mixture into a tall glass of fresh milk, with lightning speed, before a multinational panel of judges, who awarded them points in the categories of speed, swirl size, dissolve, beverage containment, presentation, and degree of difficulty.

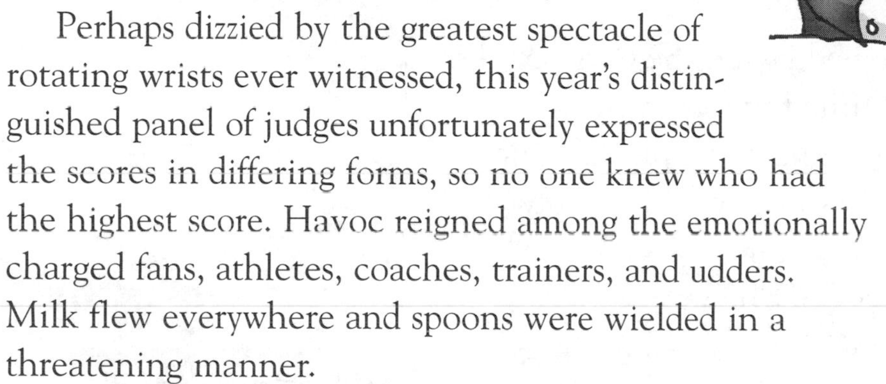

Perhaps dizzied by the greatest spectacle of rotating wrists ever witnessed, this year's distinguished panel of judges unfortunately expressed the scores in differing forms, so no one knew who had the highest score. Havoc reigned among the emotionally charged fans, athletes, coaches, trainers, and udders. Milk flew everywhere and spoons were wielded in a threatening manner.

Please return the brotherhood and humanity to the once stirring sport of stirring by ordering these scores from least to greatest:

25% $\frac{1}{8}$ $\frac{4}{5}$ .66 -3.2 -32 5% $\frac{7}{2}$ $\frac{11}{12}$

11 Number Sense & Operations

Problem

A High Chair Full of Brains

Bartholemew's mother, now better known as the mother of Bartholemew the Baby Genius, first discovered his remarkable abilities while he sat in his high chair, learning to spoon-feed himself a bowl of unrecognizably squished squash from a baby-food jar. At first, he seemed like any other high-chair occupant. He'd get a glob of food on his tiny spoon, open his mouth wide, and stick it in his eye, in his hair, up his nose, or on his belly. About one in five bites went in the neighborhood of his mouth.

On this particular day he had begun the ever popular game of throwing his squash-filled spoon off his tray onto the floor and wailing until his mom picked it up. His mother could have harvested the world's supply of coffee beans in the number of times she bent over and rose up with that spoon. She was clutching her back, dripping with sweat, and hanging off the tray. Bartholemew looked great. Squash was a good color on him.

Bartholemew's mother finally asked weakly, "Bartholemew, sweetie, how many times are you going to throw your little spoonie?"

Clear as a bell the baby responded, "$5^4 \times 5^2$ times."

His mother gasped in amazement. Then she looked puzzled. "Is that

a. 5^8, b. 5^6, c. 25^8, d. 25^6, or e. 25^{16},

honey pie?"

"How do I know? What do you want from me? I'm a baby," Bartholemew answered.

"Of course," said his mom. "It's just that if it's e. 25^{16}, I think I'll have to get a sitter."

So, which is it?

For Love of Money

As often happens, money had created false friends and obstacles to peace in the long life of Snedley Measlebop. He was so rich, his hamster had a personal trainer. Puffy could run on the gold wheel by himself, but he needed someone to hold his feet for his sit-ups.

Snedley's overly indulged forty-five-year-old nephew Philip had talked Snedley into leaving him money to invest in a leisure school, where, for an exorbitant tuition, rich young men and women could study the fine art of doing nothing. They'd be trained in holding a book so it would look like they were reading and saying, "I'm bored," in the official languages of six different countries where there certainly is plenty to do. One could receive accreditation in remote control or hip shifting (the science of lounging without losing feeling in your body parts). Fortunately, just before he died,

Snedley had a change of heart and amended his will. It now read:

> The original 4.3^6 million I was to leave my nephew Philip should be divided by 4.3^2, which he may have if he can express the quotient as a power of 4.3. He has ten seconds to get the answer. Otherwise, my attorney is instructed to donate the full value of my estate to ambitious programs to improve the lives of the disadvantaged.

Can you do it—can you explain the answer?

If so, don't tell Philip.

13 Number Sense & Operations

Problem

Couldn't Be Nicer

I don't want to say anything bad about Shelly. Shelly is a nice guy, but he sometimes forgets to use reason. He bought cereal to get the toy inside until a later age than people usually do. Most adult men don't wear Mr. Zappo decoder rings. He once bought cheese for his computer mouse, and he was surprised when his neighbor's Kerry blue terrier was not actually blue. He's a nice guy, though. He's honest. He's always willing to help. I love the guy, but he thinks $\frac{3}{4} + \frac{4}{5}$ is equal to $\frac{7}{9}$. Can you think of a clear way to nicely explain to him why this is generally not true?

Aunt Sally to the Rescue

PLEASE	EXCUSE	MY	DEAR	AUNT	SALLY
A	X	U	I	D	U
R	P	L	V	D	B
E	O	T	I	I	T
N	N	I	S	T	R
T	E	P	I	I	A
H	N	L	O	O	C
E	T	I	N	N	T
S	S	C			I
E		A			O
S		T			N
		I			
		O			
		N			

Please excuse my dear Aunt Sally.
She often does stuff out of order.
Where others start out very small,
She was tall and then grew shorter.

Please excuse my dear Aunt Sally.
She often leaps before she looks
And reads from ends to beginnings of books.

Please excuse my dear Aunt Sally.
She may drip while helping with math.
But please excuse my dear Aunt Sally.
She towels off before her bath.

There is one place she gets it right.
The order she brings to operations.
She'll make your math problems quick and easy,
Leaving time for more fun on vacations.

Using the order of operations, find the value of the following expressions with the help of dear Aunt Sally.

A. $2 + 6 * 4 - 1$

B. $3 + 4^2 - 6 * 5$

Everyone Is Wearing It

Tonight is the night of the highly anticipated fashion show of designer Jacques Flock, Paris' hottest fashion designer of the last century. Jacques is working in plastic this year. He has made bright green plastic suits and fire engine red plastic ball gowns. Backstage there is the typical chaos. Makeup artists are penciling facial expressions onto models who are scooched, wrapped, taped, and pinned into striking Jacques Flock wearable art. The lights are hot. The cloud of hairspray won't break up until it's midway over the Atlantic Ocean.

Jacques Flock, wearing a peacock blue plastic tuxedo with tails, walks through the curtain and down the runway to the *oohs* and *ahhs* of an excited crowd. He steps up to the microphone and effuses, "I'm so hot! I'm so hot! I'm so hot!"

He begins to weep. The crowd cheers. He steps behind the curtain and casts a critical eye over the numbered lineup of models before they explode onto the runway with an attitude that insists that everybody who's anybody needs to wear exactly what they are wearing. Uh-oh, there are not enough models. You have to jump in. Thank goodness you're there. You're dressed in a tangerine orange plastic dress before you know it. Now you know why Jacques was crying. It's hot inside plastic clothes. There's a zip-top bag on the side of your dress for vegetable storage and the garment comes with a bicycle pump to inflate certain sections. Still, it's better than the high-heeled flippers the guy beside you is wearing.

"When should I go out?" you ask Jacques.

"You, my tangerine surprise, are $\sqrt{51}$," he shouts.

What two integers are you in between? Just out of curiosity, which are you closer to?

Extra! Extra!

What if it turns out you don't ever need this kind of math? Then this might be your last chance to do it. Let's do some more.

1. Give the prime factorization of 36.

2. Find the greatest common factor and least common multiple of 8 and 12.

3. What is the greatest prime number less than 100?

4. Determine whether 3,456 is divisible by 3, using the Divisibility by 3 Rule.

5. Add, without a calculator:

$$3.45 + 76.421$$

6. Multiply, without a calculator:

$$3.41 \times 2.3$$

7. Divide, without a calculator, to the nearest thousandth:

$$34.6 \div 5.4$$

8. Subtract, without a calculator:

$$\frac{5}{12} - \frac{2}{9}$$

9. **Order the following set of numbers from least to greatest, without using a calculator:**

$1.3 \quad \frac{5}{7} \quad \frac{6}{5} \quad \sqrt{5} \quad 80\% \quad {}^{-}1 \quad {}^{-}1.5$

10. **Find x:**

$$3^x * 3^5 = 3^{12}$$

11. **What is $6^7 \div 6^5$?**

12. **Using the order of operations, find the value of $3 * 4^2 + (7 - 11) \div 4$.**

13. **Between which two integers is $\sqrt{13}$?**

Teacher Notes

1. A Few Good Factorers

In problem 1, students are asked to give the prime factorization of 56. To do this, they must write 56 as a product of prime numbers. Some students may use a factor tree, first finding any factor pair of 56 (for example, 7 and 8) and then continuing to factor any composite numbers (numbers with more than two factors) in the factor pair. Some students find it helpful to circle the prime numbers in their factor tree, so that in writing the prime factorization they include only the circled numbers.

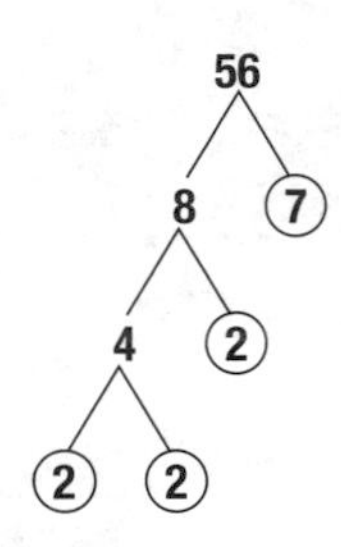

The prime factorization of 56, then, is $2 \times 2 \times 2 \times 7$ or $2^3 \times 7$. Knowing the prime factorization of numbers is a useful tool in finding common denominators when students work with adding and subtracting fractions.

2. Lovers and Other Numbers

For problem 2, students first need to find the greatest common factor of 9 and 12. The phrase *greatest common factor* is often confusing for students, given the order of the words. Since *greatest* is first in the phrase, it would be easy to think incorrectly that we're looking for numbers larger than the number given. It is sometimes helpful to think of the phrase backward: the factor common to both numbers that is the greatest. Students should first find the factors of each number. The factors of 9 are 1, 3, and 9. The factors of 12 are 1, 2, 3, 4, 6, and 12. Therefore the greatest factor common to both is 3.

Similarly, *least common multiple* may confuse students. Again it is helpful to think of the phrase backward: the multiple common to both numbers that is the least. It is a good idea to list some multiples of each number first and then compare them. The first four multiples of 9 are 9, 18, 36, and 45. The first four multiples of 12 are 12, 24, 36, and 48. Therefore the least common multiple of 9 and 12 is 36.

Another effective model for finding both the greatest common factor and the least common multiple is to draw a Venn diagram of the prime factors of each number.

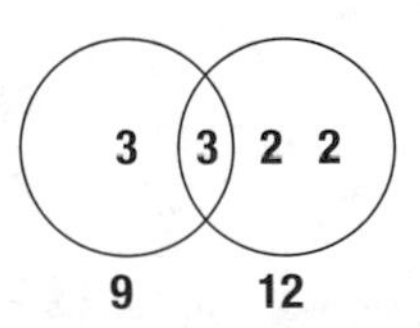

Note that a factor shared by both 9 and 12 is 3, and the 3 appears in the intersection of the two circles. To find the least common multiple, simply multiply the numbers that are in each section of the diagram: $3 \times 3 \times 2 \times 2 = 36$. Again, these concepts will be useful when students add and subtract fractions and when they factor polynomials in later grades. As an extension, students may be interested in knowing that the *proper factors* of a number are all of its factors except the number itself. For example, the *proper factors* of 6 are 1, 2, and 3. A *perfect number* is one whose proper factors add to the number. Therefore, 6 is a perfect number since 1 + 2 + 3 = 6. Can your students find a perfect number between 20 and 30?

3. The Unparalleled Thrill of Prime Numbers

For problem 3, students need to find the next prime number after 19, which is 23. Numbers that are not prime are called *composite*. (1 is the exception, since it is neither prime nor composite.) 20, 21, and 22 are all composite, since they have factors other than 1 and themselves. One interesting application of prime numbers involves the use of computers to calculate larger and larger prime numbers. One particular class of primes is the Mersenne primes, prime numbers that can be expressed as a power of 2, minus 1. For example, 31 is a Mersenne prime, since it is $2^5 - 1$. As a challenge, ask students to find the smallest Mersenne prime.

4. Divide or Be Devoured

For problem 4, students need to know the Divisibility Rule for 9: a number is divisible by 9 if the sum of its digits is divisible by 9. The sum of the digits in 2,123,456 is 23, which is not divisible by 9. Therefore 2,123,456 is not divisible by 9. If students know divisibility rules, they will save time in determining prime factorizations and greatest common factors of numbers. As an extension, ask students to research the Divisibility Rule for 11.

5. Put Out the Fire-Bellied Toads

In problem 5, students are asked to work on adding and subtracting decimals within the context of the story about Thomas. One of the skills needed to answer the parts of the problem is extracting the appropriate information from the text. For example, in part A, students will probably find it useful to make a drawing of the situation, including the table the tank is sitting on and the tank itself.

Since the distance from the ceiling to the floor is 210 cm, the tank itself is 30 cm, and the distance from the top of the tank to the ceiling is 103.7 cm, students will be able to find the height of the table by calculating 210 − (103.7 + 30). The answer is 76.3 cm.

The calculation requires decimal addition and subtraction. If students are at the stage where they need beginning instruction on these operations with decimals, teachers may find it useful to start with concrete models of decimal numbers. Base ten blocks are useful manipulatives to illustrate decimals. When using base ten blocks for decimal problems, it is important to remind students that they will have to rename the components: the flats become ones, the longs become tenths, and the units (singles, or small cubes) become hundredths. However, in this case, it may be complicated to represent 103.7 using the base ten blocks (7 longs, 3 flats, 0 large cubes, and 1 of the shape that is next in sequence after the large cube—a large long!). Therefore, it may be helpful to begin by modeling the problem of 3.7 + 30 using the base ten blocks and then add 100 mentally.

Another manipulative that works well to show addition of decimals is Digi-Blocks. Since each decimal part is constructed to be one-tenth of the previous size, there is no need to rename the parts. Using Digi-Blocks, students can model this addition quite easily, using 1 block of 100, 3 single blocks, and 7 tenth blocks for 103.7. They would then add 3 ten blocks (to add 30) and arrive at the answer of 133.7. To subtract this quantity from 210 (to obtain the answer for part A), students would have to decompose one of the single blocks into 10 tenth blocks. The answer is 76.3 cm.

Another method students might choose is one that involves shading portions of 10-by-10 grids to represent the decimals. Again, this representation is convenient if you are beginning to teach decimals without larger whole numbers (e.g., .85 − .4).

If students are familiar with the algorithm for adding and subtracting decimals, they may need to be reminded that they can use the same strategies for operations on decimals that they use for whole numbers: before adding or subtracting, it is important to align the numbers so that equal place values are recorded in the same column. For example, when adding 103.7 + 30, they must line up the digit 3 (tens place value) in 30 with the digit 0 (tens place value) in 103.7:

$$\begin{array}{r} 103.70 \\ +\ \ 30 \\ \hline \end{array}$$

At times it may be necessary to add a zero to match the decimal numbers exactly. For example, students may find it useful to rewrite this addition problem as 103.7 + 30.0.

To find the height of Thomas' karate kick in part B, students will have to find the distance from the floor to the top of the tank. Since the height of the table calculated in part A is 76.3 cm, and the tank is 30 cm tall, the distance between the cover of the tank and the floor is 76.3 + 30, or 106.3 cm. Students can use methods of addition similar to those described for part A.

Part C requires students to divide $7.50 by $1.50. Some students may be at a stage where repeated subtraction is a preferred method. They would subtract $1.50 from $7.50 and keep repeating the process until they exhausted the $7.50, or got to 0. However, in grades 6 and 7, students should know an algorithm for division. One method that is particularly useful is to change the decimal division to division with whole numbers. For example, when calculating 7.50 ÷ 1.50, we can multiply the numerator and denominator by 100: $\frac{(7.50 \times 100)}{(1.50 \times 100)}$, or $\frac{750}{150}$. Now we are left with a whole number division problem, which should be comfortable for students. The answer is 5 weeks.

In part D, the fraction of toads that did not escape was $\frac{4}{5}$. To convert this fraction to a decimal, students can divide the denominator, 5, into the numerator, 4, to get .8.

6. A Brief Tale of a Long Tail

For part A of problem 6, students need to find 15.68 − 15.573, or .107. They can use methods similar to those described in the notes for problem 5. Cal's brother's tail was longer.

For part B, students need to round 31.372 to the nearest hundredth. To round to the nearest hundredth, we need to look at the digit in the thousandths place, 2. Since 2 is less than 5, the digit in the hundredths place stays the same. Therefore, the answer is 31.37.

For part C, because we are rounding to the nearest tenth, we look at the digit in the hundredths place, 7, and since it is greater than 5, we round the 3 in the tenths place up. The answer is 31.4. This would probably not make Cal feel better about his tail since it is a larger number than the original measurement of 31.372 centimeters.

Part D requires students to multiply two decimal numbers: 5.3 × 4.5. This answer tells us how much Cal's brother's tail grew in the first 4.5 weeks, and then we would add that product to the length of his tail at birth (15.68) to find the length of his tail at 4.5 weeks. To multiply decimals, students may find it helpful first to think of the decimal numbers as fractions: 5.3 = $\frac{53}{10}$, and 4.5 = $\frac{45}{10}$. Multiplying these two fractions is actually quite simple: multiply their numerators and their denominators: $\frac{53}{10} \times \frac{45}{10} = \frac{2,385}{100}$. Students should realize that when a number is divided by 100, we move the decimal point two places to the left. Therefore the answer to the multiplication problem is 23.85 centimeters. Some students may choose to ignore the decimal places initially and multiply 53 × 45, which equals 2,385. Looking at the original numbers, they can estimate that the product should be in the twenties, and therefore it would make sense to locate the decimal after the 3 to get 23.85. Of course, if students know (and understand) the traditional algorithm for multiplication of decimals, they can use it. The length of Cal's brother's tail at 4.5 weeks was 15.68 + 23.85, or 39.53 centimeters. For a challenge question, ask if your students could tell you, before multiplying, where the decimal would be located for 46.89 × 245.875.

7. Flying Cake

For problem 7, students must add $\frac{3}{7} + \frac{5}{7}$ to get $\frac{8}{7}$. Since this is a fraction greater than 1, the correct answer is (c). For a challenge question, ask students to represent $\frac{8}{7}$ as a percent.

8. Daredevil Dresses

For problem 8, students need to find the result of $\frac{5}{6} - \frac{3}{8}$. This requires that they find a common denominator before subtracting. The lowest common denominator of 6 and 8 is 24 (note that this is also the least common multiple). Therefore we can rewrite $\frac{5}{6}$ as $\frac{5}{6} \times \frac{4}{4}$, or $\frac{20}{24}$. We can then rewrite $\frac{3}{8}$ as $\frac{3}{8} \times \frac{3}{3}$, or $\frac{9}{24}$. Therefore

$$\frac{5}{6} - \frac{3}{8} = \frac{20}{24} - \frac{9}{24} = \frac{11}{24}.$$

9. Addition Prescription

For problem 9, students need to add fractions with unlike denominators, and the process is similar to that used to solve problem 8, where they needed to find a common denominator. In this case, students have to find a common denominator for 12 and 5. The least common denominator is 60, since 12 and 5 are relatively prime (they have no common factors). Therefore

$$\frac{5}{12} + \frac{3}{5} = \frac{25}{60} + \frac{36}{60} = \frac{61}{60} = 1\frac{1}{60}$$

10. The World's Biggest Mix-Up

In problem 10, students are asked to order a set of numbers from least to greatest. Encourage students to use number sense to do a preliminary ordering: a first sorting might involve grouping the negative numbers together as the least numbers and then grouping the positive numbers.

Students should recognize that $^-32$ is less than $^-3.2$. They should also see that there is only one number greater than 1, which is $\frac{7}{2}$, so this must be the greatest number. After that, it is often helpful to put the numbers in the same form. If we make them all into fractions, the positive numbers become the following:

.25%	$\frac{25}{100}$
$\frac{1}{8}$	$\frac{1}{8}$
$\frac{4}{5}$	$\frac{4}{5}$
.66	$\frac{66}{100}$
5%	$\frac{5}{100}$
$\frac{7}{2}$	$\frac{7}{2}$
$\frac{11}{12}$	$\frac{11}{12}$

Some students may be able to order these numbers in this form; others may need to find a common denominator to compare them. If students recognize that $\frac{11}{12}$ is larger than $\frac{4}{5}$, and that $\frac{5}{100}$ is smaller than $\frac{1}{8}$ (which is $\frac{12.5}{100}$), then they can put the numbers in order.

If we write all the numbers in decimal form, however, the comparison is much easier. The decimal forms are

.25%	.25%
$\frac{1}{8}$	.125
$\frac{4}{5}$	.8
.66	.66
5%	.05
$\frac{7}{2}$	3.5
$\frac{11}{12}$	$.91\overline{6}$

From least to greatest the numbers are

$$-32,\ -3.2,\ 5\%,\ \frac{1}{8},\ 25\%,\ .66,\ \frac{4}{5},\ \frac{11}{12},\ \frac{7}{12}$$

In general, decimals are much easier to compare than fractions. It is important for students to recognize and become facile with moving from one form of a rational number to another (converting among decimals, percents, and fractions). Depending upon the context, some forms are easier to use than others. As an extension problem, ask students where they would place 120% in the list of numbers.

11. A High Chair Full of Brains

Problem 11 deals with multiplying numbers with exponents. If students do not remember the laws of exponents, you can ask them to think about the meaning of 5^4 and 5^2.

$$5^4 \text{ means } 5 \times 5 \times 5 \times 5$$

$$5^2 \text{ means } 5 \times 5$$

Therefore, $5^4 \times 5^2$ means

$$(5 \times 5 \times 5 \times 5) \times (5 \times 5) = 5^6$$

The correct answer, then, is (b). Some students may understand that in general, $a^b \times a^c = a^{b+c}$ (this is the product law of exponents). As an extension, ask them if $2^5 + 2^5 = 2^{10}$.

12. For Love of Money

Problem 12 is similar to problem 11 in that it requires students to think about operations with exponents. We can think of this problem in the same way we thought about multiplication with exponents: Rewrite the problem $\frac{4.3^6}{4.3^2}$ as

$$\frac{(4.3 \times 4.3 \times 4.3 \times 4.3 \times 4.3 \times 4.3)}{4.3 \times 4.3}$$

Now consider grouping one of the 4.3s in the numerator with one in the denominator, $\frac{4.3}{4.3}$, a form of 1. We can do this with both of the 4.3s in the denominator, and therefore our problem becomes

$$(4.3 \times 4.3 \times 4.3 \times 4.3) \times 1 \times 1 = 4.3^4$$

Students may know the general quotient law of exponents: $a^b \div a^c = a^{b-c}$, where a does not equal 0. As a challenge question, ask students what 4.3^0 is. The laws of exponents will be important tools for students throughout algebra. One application is in the area of scientific notation, which is used to express very large and very small numbers in notation that uses powers of ten; for example, 5.4×10^{35}.

13. Couldn't Be Nicer

Problem 13 provides an example that is often a misconception for students: when you add fractions, why not just add the numerators and the denominators? Life certainly would be easier if this were the correct procedure. After all, when multiplying fractions, we multiply the numerators and the denominators. With addition and subtraction, however, it's necessary to find a common denominator in most contexts.

For this problem, encourage students to estimate an answer before they do the actual addition. Would it be possible to add $\frac{3}{4}$ (a number close to 1) and $\frac{4}{5}$ (a number also close to 1) and get $\frac{7}{9}$, which is less than 1? If students know the decimal equivalents of $\frac{3}{4}$ and $\frac{4}{5}$ (which they should know without calculating), then they should see immediately that adding numerators and denominators does not make sense.

Note, however, that the question says, "this is generally not true." Why *generally*? In fact, when thinking about these fractions as ratios, it *would* make sense to add numerators and denominators. Consider this example: Tim gets 3 hits out of 4 in the first game and 4 hits out of 5 in the second game. What is the total number of hits per at bat for the two games? Clearly the answer is 7 out of 9.

For problem 13, however, the intent is to get students to think about adding these numbers as divided quantities. As an extension, ask students to come up with two fractions whose sum is $\frac{7}{9}$, neither of which has a denominator of 9.

14. Aunt Sally to the Rescue

For problem 14, students need to know the convention of order of operations: expressions should be evaluated in a particular order. First, calculate any expressions inside parentheses. Second, evaluate any expressions with exponents. Third, calculate multiplications and divisions from left to right. Fourth, calculate additions and subtractions from left to right. Some students find it helpful to use the mnemonic device *Please Excuse My Dear Aunt Sally* (PEMDAS) to remember the order.

For part A $(2 + 6 * 4 - 1)$, therefore, we would multiply $6 * 4$ first, then add 2, and then subtract 1. The answer is 25.

For part B $(3 + 4^2 - 6 * 5)$, we square 4 first, then multiply $6 * 5$. We obtain then $3 + 16 - 30$, which is $19 - 30$, or $^-11$.

The order of operations is a convention that avoids ambiguity in calculations, and one that will be useful throughout students' mathematical experiences. As an extension question, ask students to use the order of operations to evaluate $3(2-5)^2 + 10 \div 5 * 2$.

15. Everyone Is Wearing It

For problem 15, students need to understand the meaning of the square root of a number: the square root of a number can be thought of as the side of a square with that number as the square's area. For example, the square root of 36 is the side of a square whose area is 36, or 6:

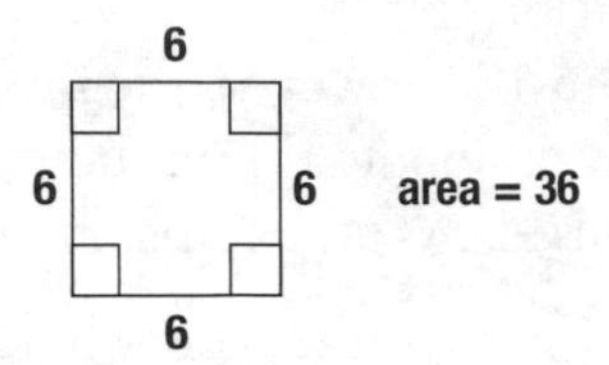

In evaluating $\sqrt{51}$, the easiest approach is to find two perfect squares, one lesser and one greater, that are closest to 51. Those numbers are 49 and 64. Therefore, $\sqrt{51}$ must be between $\sqrt{49}$ and $\sqrt{64}$, or between 7 and 8. Since 51 is closer to 49 than 64, $\sqrt{51}$ is closer to 7. As a challenge, ask students to estimate $\sqrt{51}$ to the nearest tenth.

Extra! Extra!

1. $2^2 \times 3^2$
2. GCF: 4; LCM: 24
3. 97
4. Yes, it is divisible by 3.
5. 79.871
6. 7.843
7. 6.407
8. $\frac{7}{36}$
9. $^-1.5$, $^-1$, $\frac{5}{7}$, 80%, $\frac{6}{5}$, 1.3, $\sqrt{5}$
10. $x = 7$
11. 6^2
12. 47
13. 3 and 4

Patterns, Relations & Algebra

The problems that follow are in the Patterns, Relations, and Algebra strand. The mathematics in these problems focuses on extending sixth and seventh graders' ability to describe, generalize, and extend numerical, tabular, and geometric patterns using algebraic symbols. In addition, students set up and solve linear equations.

The topics covered in these problems were chosen from state and national standards:

- Represent and analyze a variety of patterns with tables, graphs, words, and symbolic expressions
- Solve linear equations

Patterns, Relations & Algebra

Problem

The Family That Longed to Pass the Peas

Although they are wonderful children, my son, Thomas E, and my daughters, Alley and Toshia, fight constantly over legroom through every meal. We have a very small kitchen table. World Cup soccer doesn't have this much kicking. As I pour the milk, I call them in to wash their hands and strap on their shin guards for dinner.

There's no room for me at the table. I sit on a step stool in the corner, kind of like Cinderella, but without the rich fantasy life. I never eat more than a bite or two at a time before I have to bring another serving to the battling children because the table is too small to keep our serving dishes on the table. They finish their servings quickly because we use dollhouse dishes from Alley's "Baby Sit 'n' Spit" that she had when she was little. It's a very small table.

Here's a diagram of our tabletop.

Sometimes, exhausted from toting plates heaped with two or three beans, I dream of having the tabletop enlarged, like this

or even this.

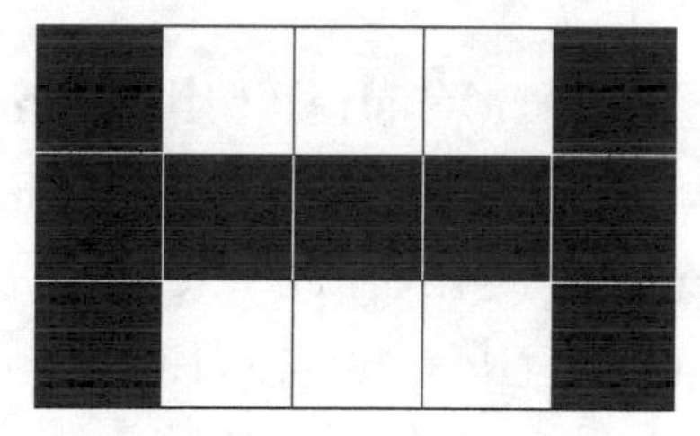

I talked to Zana, an interior designer from the Too Nice to Sit On furniture boutique, about having the table enlarged but still keeping the black-and-white tile pattern. She said that eggplant-colored tiles are the rage and that normally they are very, very expensive, but, because she likes me, she could get some for me that were just *very* expensive. I thanked her for liking me but said that I really just wanted the continued black-and-white tile pattern. She made a face as though she was remembering finding a rotten banana in her purse and asked me if my son's bedroom had a theme.

"Sure," I said, "dog hair and broken toys."

"How nice," she said. But I don't think she really thought it was nice.

See the pattern in the table diagrams of the H-shaped figures? Please help me explain it to Zana before she tries to redo my living room.

A. Make a table of the number of white square tiles and the number of black square tiles for the first three stages of the pattern. (For example, in the first stage, shown on the previous page, the number of white tiles is 2, and the number of black tiles is 7.)

B. Find a rule that links the number of black tiles to the number of white tiles.

C. How many black tiles would there be for 8 white tiles?

D. You're doing great, but Zana's still not getting it. Plus, now she's talking about putting skirts on the kitchen chairs, which I'm sure would look lovely, but I don't like my furniture dressed better than me. Could you make a graph of the first four pairs of numbers for this pattern? Put the number of white tiles on the horizontal axis (x-axis) and the number of black tiles on the vertical axis (y-axis).

E. Do the points lie on a line?

F. Make another table of the number of black tiles and the total number of tiles for the first three stages. For example, in the first stage, there are 7 black tiles and 9 total tiles.

G. Can you find a rule that links the total number of tiles to the number of black tiles?

H. How many total tiles would there be for the stage where there are 20 black tiles?

Kitchen Disasters

The only thing I can do successfully in the kitchen is rinse out the sink. I can't cook or bake. It makes no sense because I have measuring spoons, I can turn on the stove, any fool can mix, and I'm an expert at "set aside," which is often an instruction after one set of ingredients have been combined. I can do each step, just not in a row. It's like how I rumba.

I removed the cardboard from the bottom of the pizza the other night after I cooked it. The pizza's not as good that way. The dough doesn't cook and it takes a long time to get the cardboard off. No matter how much a cookie recipe makes, I burn, drop, disfigure, or otherwise make inedible 12 of them before I can set them out on a decorative tray. Today I quintupled (multiplied by 5) the oatmeal raisin cookie recipe and ended up with 23 cookies. How many cookies did the original recipe make? Here's the problem:

$$5x - 12 = 23$$

You figure it out. I'm rinsing out the sink.

One Potato, Two Potatoes

For dinner last night I made baked potatoes. I forgot what temperature to cook them at and I don't own a cookbook. I don't think there's a recipe for baked potatoes anyways. I think people are just born into the world knowing how to bake potatoes. I just closed my eyes and turned the temperature dial on the oven. I figured that if it was too hot, the potatoes would say something. It didn't matter anyways; I was in a hurry and didn't really have time to leave the potatoes in there very long.

They came out cooked on the outside, raw on the inside, but not unhappy. When we cut them at the table, the knife would go easily through the top and then get stuck. When we pushed down on the knife, a chunk of potato would go flying off the plate.

There were four potatoes. Seven pieces of potato and I don't know how many cheese crumbles went flying from each. The cheese crumbles were mild cheddar, but nothing is very mild when it's rocketing toward

HEY!

your head without control. I personally spilled three dollops of sour cream. I've picked up less food from a busted pinata. There were fifty-one food items on the floor after the root rumble.

How many cheese crumbles did each potato fire off?

I've written the problem here as a linear equation.

$$4(x + 7) + 3 = 51$$

Extra! Extra!

Many people never get the chance to do extra patterns, relations, and/or algebra problems. You're not one of them.

1. Give the following sequence of figures:

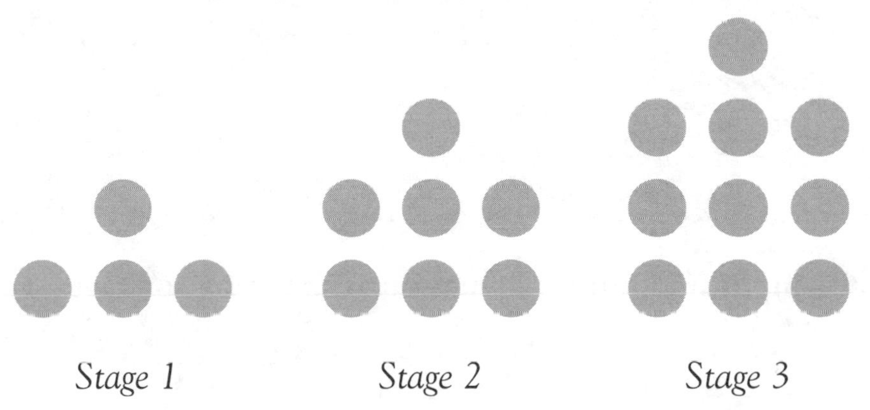

A. Make a table of the number of circles in each stage, with the input being the stage number and the output being the total number of circles.

B. Find a rule that relates the number of circles in each stage with the stage number.

C. How many circles would be in stage 100?

D. Graph the first 4 input and output values corresponding to stages 1, 2, 3, and 4.

E. Do the points lie on a line?

F. What stage has 127 circles in it?

2. Solve the following equations:

A. $6x - 13 = 24$

B. $3(x - 5) + 2 = 20$

Teacher Notes

1. The Family That Longed to Pass the Peas

In part A of problem 1, students are asked to create a table of values for the number of white and black square tiles in each stage of the pattern. The table should look like this:

NUMBER OF WHITE TILES	NUMBER OF BLACK TILES
2	7
4	8
6	9

For part B, students need to find a rule that relates the number of white tiles to the number of black tiles.

Students may first notice that there are always 3 black tiles on each side of the H figure. Also, the number of additional black tiles—the tiles that form the horizontal part of the H—is always half the number of white tiles. Therefore, the rule is: $y = \frac{1}{2}x + 6$, where y is the number of black tiles, and x is the number of white tiles.

Another way to think about the problem is to focus on the change in values in the table, looking for constant differences. The constant difference in black tile values is 1. The constant difference in white tile values is 2. Since these differences are constant, the relationship is linear, and the slope of the line connecting these points is $\frac{1}{2}$. Therefore the rule must be of the form $y = \frac{1}{2}x + b$. Substituting one of the points in this equation, for example (2, 7), we obtain the equation $7 = \frac{1}{2}(2) + b$, and therefore $b = 6$. Again we discover that the equation is $y = \frac{1}{2}x + 6$. Even if students use the tabular approach in the second explanation, it is often helpful for them to see where the numbers $\frac{1}{2}$ and 6 come from in the context of the tile pictures.

For part C, to find the number of black tiles for 8 white tiles, we can look at the picture and notice that for 8 white tiles there would be half as many black tiles in the horizontal part of the H (4 black tiles) in addition to the 6 tiles on the vertical portion of the H, for a total of 10 black tiles. If we use the equation, $y = \frac{1}{2}x + 6$, and substitute 8 for x, we obtain $\frac{1}{2}(8) + 6$, or 10.

In part D, students are asked to graph the first 4 pairs of numbers for the pattern. The graph would be:

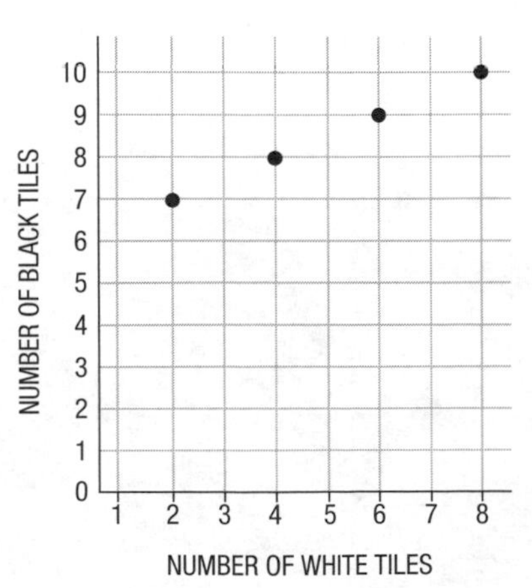

In answer to part E, the points do lie on a line. This should make sense intuitively to students since there is a constant increase of 1 black tile for every increase of 2 white tiles. As an extension, ask students if it would be appropriate to connect the points on this line.

For part F, the table showing the number of black tiles and the total number of tiles should look like this:

NUMBER OF BLACK TILES	TOTAL NUMBER OF TILES
7	9
8	12
9	15

We can derive the rule asked for in part G by looking at how many tiles are added from one stage to the next. There are always two additional white tiles and one additional black tile, for a total of 3 tiles. Therefore the constant rate of change is 3, and the rule is $y = 3x + b$. To find b, take one of the points in the table, such as (7, 9), and substitute those values into the equation: $9 = 3(7) + b$. Solving for b, we find $b = {}^{-}12$. Therefore the rule is $y = 3x - 12$.

It is important that students be able to analyze the behavior of patterns from tables. In later grades, they will discover that if the second differences (the differences of the differences) are constant in a table of values, the relationship is quadratic. What do you think is true of relationships with constant third differences? (Third differences are the differences of the second differences.)

For part H, to find the number of total tiles if there are 20 black tiles, substitute 20 into the equation: $y = 3(20) - 12$. Therefore, $y = 48$. As an extension, ask students to find the number of black tiles if there are 78 total tiles.

2. Kitchen Disasters

For problem 2, students are required to solve a linear equation using any method they choose. For example, students can think of building a flowchart and working backward, or backtracking. The flowchart for this equation would be:

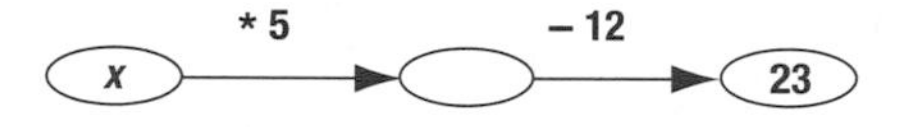

Note that x is the input, and 23 is the output. Working backward from the output, what number less 12 is 23? The number must be 35. Now, looking at the next mathematical action, what number multiplied by 5 is 35? The input number, or x, must be 7. Students begin to realize that they are using the inverse operations when they work backward: adding 12 and dividing by 5.

Some students will solve this equation by doing the same thing to both sides: first adding 12 to both sides and then dividing by 5. Note that these are the same actions as when working backward. As an extension, ask students to write out the driving directions to a friend's house and then write the reverse directions. Can they generalize reversing any directions?

3. One Potato, Two Potatoes

For problem 3, the flowchart would be

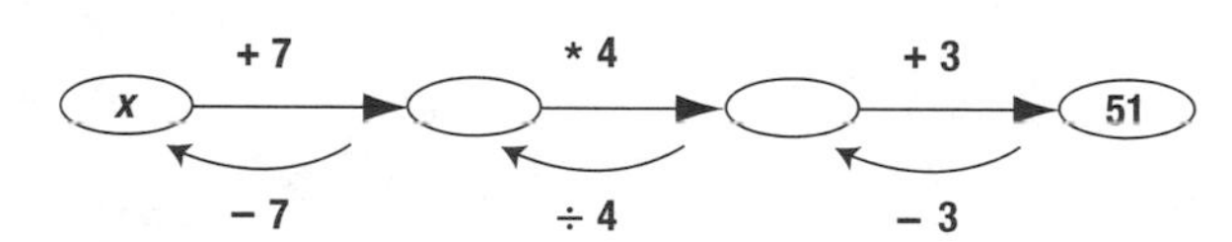

Beginning with the output, 51, subtract 3 (resulting in 48), then divide by 4 (resulting in 12), then subtract 7. The final answer is 5.

Extra! Extra!

1. A.

STAGE NUMBER	TOTAL NUMBER OF CIRCLES
1	4
2	7
3	10

B. $c = 3s + 1$, where c is the total number of circles and s is the stage number

C. 301

D.

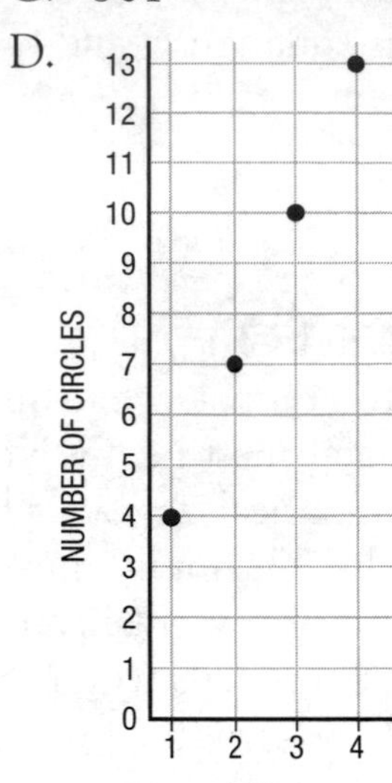

E. Yes, they lie on a line.

F. stage 42

2. A. $x = \frac{37}{6}$, or $6\frac{1}{6}$

   B. $x = 11$

Geometry

The problems that follow are in the Geometry strand. The mathematics in these problems focuses on student exploration of three-dimensional figures, the Cartesian coordinate plane, and angles formed by intersecting lines. Students in grades 6 and 7 continue to look at the area and perimeter of two-dimensional shapes and analyze the relationship between the number of sides a polygon has and the sum of its interior angles.

The topics covered in these problems were chosen from state and national standards:

- Apply concepts of perimeter and area to squares, rectangles, triangles, and circles
- Apply the relationship between the number of sides and the sums of the interior angle measures of polygons
- Understand the relationship of angles formed by intersecting lines
- Graph points on the Cartesian coordinate plane
- Identify three-dimensional figures (e.g., prisms, pyramids) by their attributes

1 Geometry

Problem

Look Out Below, Please

Mr. Bimm was a quiet, careful man who loved language, flowers, and civility. He had spent years studying and appreciating the manners and customs of different cultures. On a windy day on the balcony of his thirtieth-floor apartment, Mr. Bimm was caring for his garden. As he lowered a pale green stem into its new home in a more spacious pot, he thought, "When the Wahui tribe planted crops, they danced in the hot sun for a full day while sprinkling the perimeter with wolf urine. I think I'll just water it and talk nicely to it."

At just that moment a great gust of wind blew over an empty vase sitting on top of the railing, toppling it onto Mr. Bimm's head. There was a loud *phthunk* sound as the vase forced its way over his ears, and poor Mr. Bimm was in total darkness. He staggered back and his vase-enclosed head went over the decorative iron railing, followed by the rest of his body parts. Fortunately, he grabbed hold of a rung of the railing, but then, there he hung. "When the Wahui Indians were in danger, they sang a little song of bravery and then spit twice. I believe I'll just yell for help," he thought. And he did.

Sooner than it felt like to Mr. Bimm, a rescue team assembled on the ground beneath him. They unfurled a large canvas and shouted instructions to Mr. Bimm. "Dangling man with the vase on your head! Swing away from the building and drop," directed a rescue worker through a bullhorn.

Mr. Bimm knew that he should shout back, "Yes, thank you," but instead he shouted, "How big is your canvas?"

"It's rectangular with a 24-foot perimeter," blared the bullhorn. Two globs of spit dribbled out of Mr. Bimm's vase.

A. Find the dimensions of a rectangle whose perimeter is 24 ft.

B. What is the area of your rectangle?

C. Can you find a different rectangle with a perimeter of 24 ft.? Calculate its area as well.

2 Geometry

Problem

Leonardo Digit: The King of the Continual Curve

It's the dream of many a young person to play someday for the California Angels. However, because of a misprint on his dream form, Leonardo Digit got an opportunity to play for the California Angles, a geometry team. It's harder than it sounds. He got injured in his first game.

It's now pretty late in the season, and what a season it's been. The Angles were in a heated battle with the other top teams for the finals in big league geometry, and Leonardo had made a name for himself as the answer man on circle questions. He came through in a clinch over and over. They called him The Sultan of Circles, The Babe of the Ball Shape, and The King of the Continual Curve.

This next question could capture the coveted finals spot for the Angles. The crowd is on the edge of their seats and leaning forward at a 45-degree angle. Leonardo Digit is up. The question flies in. Leonardo touches his cap, looks left, looks right, and feels a swelling sense of panic. He's only good at circles. Jump in, Triangle Titan, and help him answer the questions that follow.

A. A triangle has a base of 8 inches and a height of 3 inches. Find its area.

B. If the triangle is isosceles, find its perimeter.

HINT. You may need to use the Pythagorean Theorem.

3 Geometry

Problem

The Breakup of Cookies and Milk

Relations had become strained between the tiny sovereign countries of Rutabaga and Rhubarb. Each country had a different view of what brought about their hostilities and you can easily guess who thought who did what. Whatever the cause, years of friendly trade and trust had come to an end, leaving both countries less than they had been.

Rutabaga's chief export having been cookies and Rhubarb's having been milk, it was an especially sad situation. Rutabaga didn't produce milk. It had never needed to with Rhubarb doing such a great job of it right across the way. And Rhubarb was too busy keeping its dairy cows happy to have ever learned to make cookies. The country had been happy to obtain the best cookies in the world from Rutabaga. Now, Rutabagans angrily choked down dry cookies while Rhubarbites filled with rage as they sucked down gallons of plain old milk so it wouldn't go sour and stink up their land.

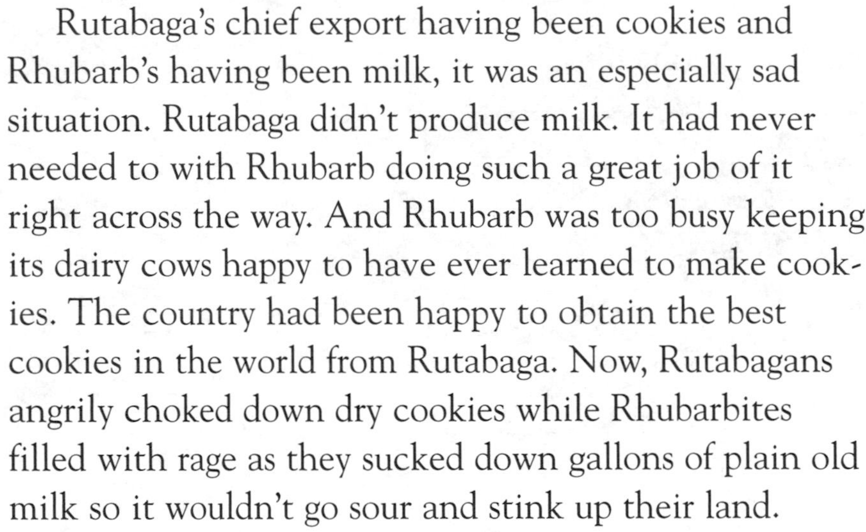

Each country formed a small army (about five soldiers each; they were small countries) to stand toe-to-toe with their enemy counterparts at the border. The Rhubarbite troops wore milk mustaches and the Rutabagan force sported cookie crumbs on the front of their uniforms to further agitate their rivals. Things were tense. Each country, certain that the other was up to no good, secretly sent spies across the border with strict instructions to contact headquarters if they saw anything unusual.

Agent LMNO from Rhubarb sat at a milk bar in a tall building in Rutabaga, staring out the window through a hidden telescope in the bottom of his empty milk glass and talking into his cheese. "LMNO to Headquarters. Over," he whispered.

"Headquarters here, LMNOP," came a voice from his earmuffs.

"I'm LMNO, not LMNOP," corrected LMNO.

"I know," said the voice from Headquarters, "I just can't say LMNO without saying P. It's how I was raised. Anyways, what do you have for us?"

"I have sighted a huge circular pool of milk with a circumference of 20 meters."

"Excellent, Agent . . . you know. Can you tell us the diameter and the area of the circle?"

"I'm afraid I can't," LMNO breathed nervously. "The waiter just refilled my telescope with milk and if I don't eat my cheese, they'll be onto me."

Headquarters heard a loud chewing noise and the line went dead. It's best for you not to take sides in these things, but . . .

A. What was the diameter of the circle?

B. What was the area of the circle?

Problem

Heads Up

"What is the significance of the five sides of the hat?" shouted the reporter for the *New York Herald.*

"Why the five sides of the hat?"

"How come the hat has so many sides?"

"¿Porqué el sombrero tiene cinco lados?"

"What made you think of five sides for a hat?"

"Have you ever been in trouble before?"

"I don't get it!"

"Five sides? What, are you crazy?"

"My heavens!"

Questions and comments like these continued to burst forth from the microphone-waving television reporters from channels 2 through 60, including the fashion channel, the religion channel, the fishing channel, and the channel for people who just sit in chairs.

A tall, thin man with his jet black hair sculpted into a large curly-Q on top of his head stood on a platform outside an upscale store in New York, soaking up the attention. Crowds of people clamored to know about the wildly popular five-sided hat and its flamboyant designer, Jacques Flock. Not since the time when King Louis XIV told everyone that they needed a different outfit each year had any fashion achieved the popularity of the five-sided hat. Not blue jeans, not flip-flops, not the bustle, not the miniskirt, not the teeny weeny bikini. Nothing had ever been as popular as the pentagonal hat. Congress even passed a law increasing the penalty for theft when the theft included a five-sided hat.

Jacques stood facing the crowd and, holding up his own five-sided hat, he laid his finger gently on a side of it. The crowd hushed as he addressed them.

"The first side of the five-sided hat represents peace," he said, and the crowd nodded their approval of so beautiful an ideal and wondered why our world never has enough of it.

"The second side of the hat," he continued, pointing to the next side, "stands for love."

The crowd breathed deeply, united in their desire for a world full of love.

"The third side of the hat," Jacques went on, "represents how hot I am. And so does the fourth and so does the fifth, because I am so hot, hot, hot!" The crowd cheered. Jacques dissolved into tears and strode from the stage.

A. What is the sum of the interior angles of a pentagon like the top of the five-sided hat?

B. Would the sum of the interior angles change if the shape of the pentagon changed?

5 Geometry

Problem

Smile Pretty for the Ambulance

Hiram Dooda made enough money as the host of television's *Prime Time*—which features a different prime number each week—to take up skiing. The show was deadly dull, but you'd never know it to look at Hiram. He had a big show-business smile and, from either sheer ego or an admirably bright outlook on life, Hiram never let on that anything was as bad as it looked.

On this particular crisp, beautiful day, Hiram imprisoned his feet in the heavy, restrictive footwear that skiers love, strapped them to two slippery slats, and rode in a wind-blown chair that dangled from a cable high above a snow-covered mountain filled with people who also thought that was a good idea. The chair delivered a smiling Hiram to the tippy top of the snowy mountain, and within seconds the wind was whistling in his ears as he sped down an icy trail. Hiram and his smile looked sharp. He could have been on the "Winter Wonderland Ski Experience" brochure until his skis spread beyond his wingspan and he disappeared behind some ponderosa pines and then reappeared, traveling at remarkable speed, bearing a bird's nest on his head that he definitely didn't have before. His eyes showed just a trace of panic, but his mouth held a big show-business smile, perhaps out of habit. Being a bit off balance, he stabbed at the air with his ski poles like a swashbuckling pirate, which was probably the only thing protecting him from the bird the size of a pteranodon that was dive-bombing his head just as angrily as you would if someone skied away with your house on his head.

News travels fast, so by the time he slid through the Mile High Refreshment Chalet and came out the other side dragging the coffee urn, crowds had begun to gather along the slopes. Some were curiosity seekers and some were angry coffee drinkers, but all got a terrific view of the first moose ever spotted on the mountain when it came charging after Hiram, who had disturbed it by flying over the rocky ledge it was resting under. "My bad," shouted Hiram, always willing to take responsibility for his mistakes.

Hiram doesn't remember any of this, but legend has it that when he finally landed in a heap after crashing into the passenger side of a ski patrol SUV in the parking lot, he regained consciousness long enough to see a sea of human faces and that of one moose peering down at him, and when he heard someone say, "That's Hiram Dooda from *Prime Time*. His holiday special about 59 was awesome! My dad and I watched it," Hiram smiled even bigger, and said, "We'll be right back with more prime number excitement," and passed out again.

When Hiram landed, his skis were crossed like so:

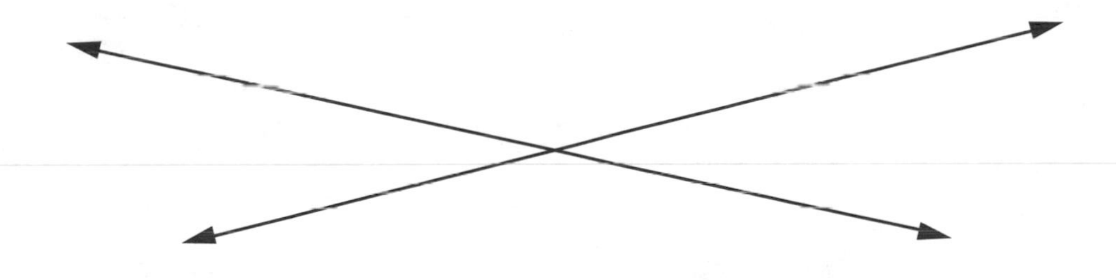

A. If one of the angles formed by Hiram's skis was 40°, find the measures of the other three angles.

B. If one of the angles was $x°$, find the measures of the other three angles.

Dead Men Do No Math

Young Armando awoke in the ship's cabin in terror, with the point of Captain Stubbly Beard's cutlass on the tip of his nose and Captain Stubbly Beard at the other end with a firm grip on the handle. "Aargh, matey, I hears you've got the coordinates of the route to the treasure. Ya best be quick about handing it over 'cause if I got no treasure, I got no patience. I'm quite patient, though, when I do have treasure. I still make them that cross me walk the plank, but I don't care how long it takes 'em. But now I got no treasure, so get your barnacles up out of that bed and give me the coordinates," Captain Stubbly Beard sneered.

"But, sir," Armando began, until Captain Stubbly Beard poised for a lunge. Armando pulled a paper from his pocket and placed it in the filthy palm of the vile captain.

"And, now, you scurvy bilge rat, get on deck," the pirate shouted.

Quickly, Stubbly Beard plotted the coordinates on a graph he had laid over a chart of the region. Then he announced, "We're changing course," charged to the helm, and, pushing aside the helmsman, began to steer.

"Raise the main sail. Jab the jib up or you'll sail from the long Tom," Stubbly Beard bellowed. And they must have done it because the ship lurched and cut through the water with tremendous speed until, staring carefully at his chart, Stubbly Beard spun the helm like a top.

The ship reeled. The sails leaned nearly parallel with the sea. Men clung to whatever they could, and still there were screams of those whose sea-soaked grip gave way, plunging them to their end in the depth of Davy Jones' locker.

"Get this tub to right, ya scurvy dogs," the captain blasted.

And they did, only to be practically capsized with each of four more deadly, abrupt turns.

Cutlass or no cutlass, Armando staggered to face the bloodthirsty captain. "Sir," he cried through the sea spray, "those coordinates are for my math homework. They've nothing to do with treasure."

The captain's eyes narrowed and the corners of his lips snuck across his face.

"You, there," he snarled, waving to a bilgewater-covered pirate holding a telescope. "Pass me the bring-'em-near."

"I think you can ask more nicely," said the pirate.

Stubbly Beard drew his sword, and a moment later the headless pirate quietly handed the captain the telescope. Peering through it, Captain Stubbly Beard found they were at the precise point where they began.

Armando is now walking the plank. Since dead men do no math but Armando's homework is still due, could you complete these problems for him?

A. What are the coordinates of the point halfway between (3, 8) and (3, 14)?

B. Give the coordinates of two points that, along with (3, 8) and (3, 14), form the vertices of a rectangle.

C. Graph these points.

D. Can you find two different additional points that will form the vertices of a rectangle along with (3, 8) and (3, 14)?

Don't tell Stubbly Beard.

Problem

Private O'Malley on the Edge

Private O'Malley's unit from the Division Division of the army is on maneuvers. They were in a group of 100 when they started marching this morning, and they divided into two groups of 50. Then O'Malley's group split into two groups of 25. His group of 25 headed straight up a mountain trail, double time. Then 4 groups of 5 cut out. Two groups went left; two groups went right. O'Malley's remaining group of 5 continued straight up the mountain. This divided his happiness. As they neared the mountain's peak, the good news was that they were almost to the top. The bad news was that, with very few trees at the top, the wind blew 4 of the 5 guys back down the mountain, leaving Private O'Malley the lucky lone survivor.

O'Malley didn't feel lucky. It was getting dark. He was cold. His nose was running faster than a track-and-field star. All of this dividing was, he was pretty sure, driving him insane. Last night he had a dream that he was being chased by a monster who caught him only to measure him so that the monster could cut him into equal parts and eat him in tomorrow night's dream. He was really sick of this. He had had it, or he had had at least $\frac{7}{8}$ of it.

He dropped his heavy backpack with a moan, pulled out his tent, and assembled it.

Here's a drawing of his triangular prism-shaped tent.

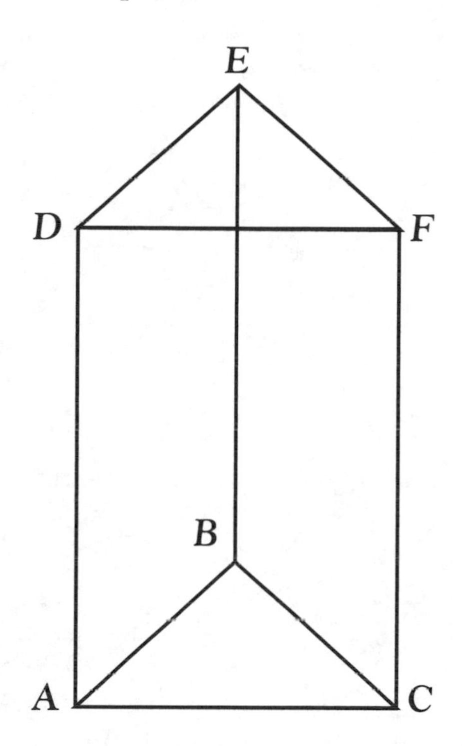

O'Malley stepped back to get a good look at his tent. "What?" he howled in despair, disturbing a flock of birds, who divided into 3 groups. "Did someone divide my tent?"

Poor O'Malley is beside himself, which is going to make it even more crowded in that tent tonight.

A. How many faces of Private O'Malley's tent are parallelograms? Name them.

B. How many bases does it have? Name them.

C. How many edges does it have? Name them.

D. The area of triangle *ABC* is 5 sq. ft., and height *FC* = 2 ft. Can you find the volume of the prism? If that were the volume of the tent, can you see why our hero is so upset?

Extra! Extra!

Here are some more geometry problems for practice. There is no end to the study of geometry. Do you see my point?

1. A rectangle has an area of 20 sq. ft. Find its perimeter. Can you find more than one answer?

2. An isosceles triangle has sides 10 cm, 10 cm, and 12 cm. Find its area.

3. Find the circumference and area of a circle with a radius of 4 inches.

4. Find the sum of the interior angles of a hexagon.

5. An isosceles triangle's base is the line segment connecting the points (5, 4) and (11, 4). If the height of the triangle is 5, find two possible coordinates for the third vertex.

6. Find the surface area and volume of a rectangular prism whose base is 5 feet by 3 feet and whose height is 7 feet.

Teacher Notes

1. Look Out Below, Please

Problem 1 requires that students know how to find the perimeter of a rectangle—double the length and width and add them together.

$$2L + 2W \text{ or } 2(L + W)$$

Many students will begin by simply picking an arbitrary number for the length or width, for example, 4. If the width is 4 and the perimeter is 24, then we can find the length by calculating $24 - (2 \times 4)$, which gives us 16. Since there are two lengths, each length must be 8. So one rectangle that can be used as a solution to part A has a width of 4 feet and a length of 8 feet.

Another way to think about the problem is to take half of the perimeter, or 12, and conclude that the sum of the length and the width is 12. Therefore, any two numbers that add to 12 could be the measures of the length and the width.

The answer to part B will vary depending on the student's rectangle. One possible response is: length of 10 ft., width of 2 ft., and area of 20 sq. ft.

For part C, another pair that adds to 12 ft. is 9 ft. and 3 ft., with an area of 27 sq. ft.

Technically, of course, if one considers all numbers as possible lengths and widths, there are an infinite number of possibilities! As an extension, you may want to ask which length-width pair would result in the largest area. Questions like "Which numbers result in the largest . . . ?" are part of a class of problems known as optimization problems, and students will encounter these throughout their mathematics careers.

2. Leonardo Digit: The King of the Continual Curve

For part A of problem 2, students need to know the formula for the area of a triangle: $\frac{1}{2}$ base $\times$ height. Note that the area for a triangle is half the area of a rectangle or parallelogram created from that triangle. (If you take a triangle and copy it, you can always make these two identical triangles into a parallelogram.) Although we do not know the exact shape of this triangle (there are an infinite number of triangles with a base of 8 inches and a height of 3 inches), we know its area is $\frac{1}{2} \times 8 \times 3$, or 12 square inches. As an extension, ask students to draw three different (not congruent) triangles with a base of 8 inches and a height of 3 inches on graph paper and observe whether the areas are the same.

For part B, students need to draw an isosceles triangle with an 8-inch base and a 3-inch height. They can do this on graph paper if it helps in drawing. Here is a picture of the triangle:

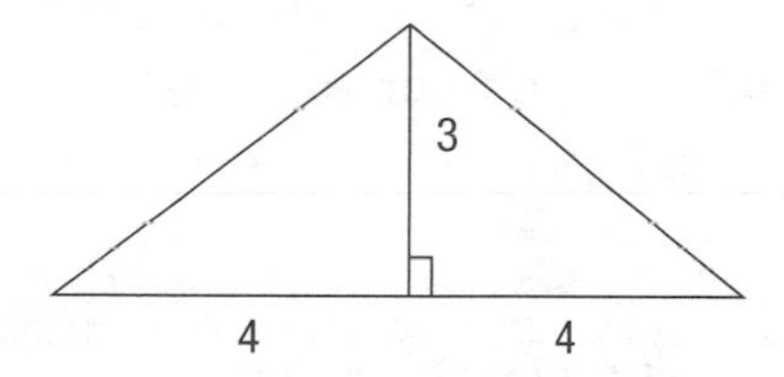

In an isosceles triangle the altitude is also a median, and the height intersects the base at its midpoint. So if the base is 8 inches, then each segment is 4 inches. To find the perimeter of the triangle, we need to find the hypotenuse of each of the smaller triangles. Since these are right triangles, we can use the Pythagorean Theorem

($a^2 + b^2 = c^2$, where c is the hypotenuse and a and b are the legs of the triangle) to find the equal sides:

$$3^2 + 4^2 = c^2$$

Solving for c, we find $25 = c^2$ and therefore $c = 5$. The perimeter of the triangle, then, is 8 + 5 + 5, or 18 inches. As an extension, ask students to find the perimeter of the triangle if it was a right triangle.

3. The Breakup of Cookies and Milk

For part A of problem 3, students need to know the relationship between circumference and diameter, which is $C = \pi d$. Therefore, if the circumference is 20 meters, then

$$20 = \pi \times d$$

$$d = \frac{20}{\pi} = 6.4 \text{ meters,}$$

rounded to the nearest tenth of a meter

For part B, to find the circle's area, we must use the area formula, $A = \pi r^2$. We know the diameter is 6.4, and the radius is half of 6.4, or 3.2 meters. Therefore the area of the circle is $\pi(3.2)^2$ or 32.2 square meters. As an extension, ask students to measure the circumference of several different round objects (they can do this with string), then measure their diameters, and then construct a table with the values for circumference, the values for diameter, and the result of dividing each circumference by its diameter. The result should be a number close to pi every time!

4. Heads Up

To find the sum of the interior angles for part A of problem 4, students should first draw any pentagon. They can divide the pentagon into triangles by beginning at one vertex and drawing segments that connect to each nonconsecutive vertex, as shown.

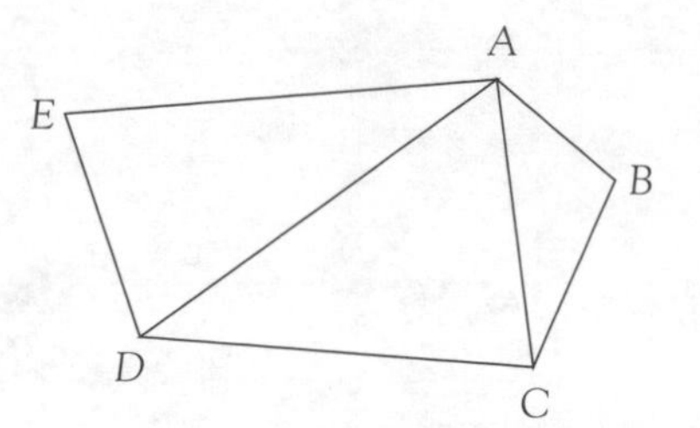

In this pentagon, we chose vertex A and connected it to vertices C and D. Note that we divided the pentagon into three triangles, and that the sum of the angles in these three triangles is the same as the sum of the angles in the pentagon. Since we know the sum of the angles in a triangle is 180°, and since there are three triangles, the sum of the angles in the pentagon is 3 × 180, or 540°.

It is helpful for students to understand that this method of dividing a polygon into triangles can be used for any polygon. As an extension, can your students find a general rule for determining the sum of the interior angles of any polygon? (Hint: They should find a relationship between the number of triangles and the number of sides the polygon has.)

As for whether the sum changes if the shape of the pentagon changes, the answer is no—regardless of the shape, the pentagon can always be subdivided into three triangles.

5. Smile Pretty for the Ambulance

To answer part A of problem 5, students need to know that angles formed by intersecting lines form vertical pairs. The pair of angles opposite each other have the same measure. They are called vertical angles or opposite angles. Although the drawing is not necessarily drawn to scale, it would appear as though the 40° angles must be in these positions.

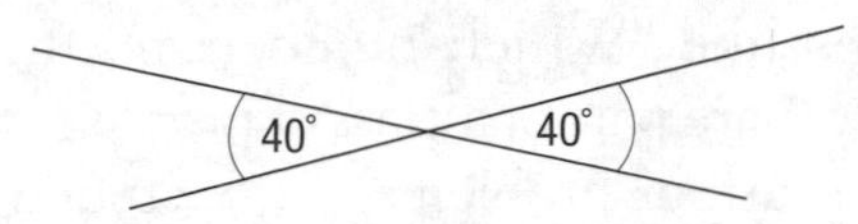

Note that the angle adjacent to either of the 40° angles forms a straight line, and therefore it and the 40° angle must be supplementary. Therefore the other angles are each 140°. The other three angles, then, are 40°, 140°, and 40°.

For part B, we can generalize the process we used in part A. If one of the angles is x, then the angle opposite it, forming the vertical pair, must also be x. The angle adjacent to it, forming a straight angle, must be $180 - x$, or the supplement of x. Therefore the measures of the other three angles are x, $180 - x$, and $180 - x$. Here's an extension for students: If two lines intersect, and one pair of angles each has a measure that's twice the angles of the other pair, find the measures of all of the angles.

6. Dead Men Do No Math

For Part A of problem 6, it might be helpful for students to plot the points (3, 8) and (3, 14) on a coordinate graph.

They will notice that the points lie on a vertical line, and that the distance between the points is 6 ($14 - 8 = 6$). Since they are interested in the midpoint of this line segment, they need to start at 8 and move 3 units (3 is half of 6) up. Therefore the point that is halfway between has the coordinates (3, 11).

For parts B and C, the graph of the two given points is a useful visual aid in determining the two additional points that would form a rectangle. Looking at the figure:

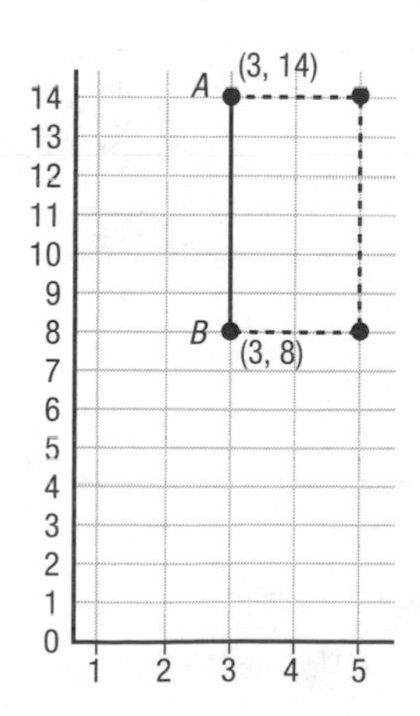

it should be clear that if AB is one side of the rectangle, the other side needs to be parallel to AB, with y-coordinates 8 and 14 as well. Therefore, one possibility is (5, 8) and (5, 14), shown in the diagram with dashed lines. In fact, in answer to part D, there are an infinite number of rectangles that can be formed, as long as the y-coordinates of the additional two points are 8 and 14, respectively. Students' ability to visualize these figures will be helpful when they study transformational geometry in the future.

7. Private O'Malley on the Edge

For part A of problem 7, students need to understand the meaning of a *face* of a prism: faces are the polygons that make up the surface of a solid. In this triangular prism, the faces that are parallelograms are *EFCB*, *EDAB*, and *DFCA*. The bases are the two end faces that are congruent and parallel to each other. In the answer to part B, there are two triangular bases (prisms are named for their bases), *DEF* and *ABC*.

The edges of a prism, which part C asks for, are the line segments formed by the intersection of any two faces; they are *AB*, *BC*, *CA*, *EF*, *ED*, *FD*, *DA*, *FC*, and *EB*.

In part D, students are asked to find the volume of the prism. The volume of any prism is the area of the base multiplied by the height. It may be helpful to think of any prism as the layering of bases: if we know the area of the base, we can think of the height as the number of layers of the base. In this case, since the area of the base, triangle *ABC*, is 5 sq. ft., and the height, *FC*, is 2 ft., the volume is 5×2, or 10 cubic feet. Note that we measure volume in cubic units. Ask your students to look around the room for an object that takes up about 10 cu. ft. Do they see now why O'Malley was so upset? As an extension, can your students find the dimensions of a different triangular prism with the same volume?

Extra! Extra!

1. One possible rectangle has dimensions of 10 ft. by 2 ft., and therefore the perimeter would be 24 ft. Another possible rectangle has dimensions 4 ft. by 5 ft., and its perimeter would be 18 feet. There are many possible answers.
2. 48 sq. cm
3. circumference: 8π in.; area: 16π sq. in.
4. $4 \times 180 = 720°$
5. (8, 9) or (8, ⁻1)
6. surface area: 142 sq. ft.; volume: 105 cu. ft.

Measurement

The problems that follow are in the Measurement strand. The mathematics includes converting between units of measurement within the same system by looking at proportional relationships. Students are also expected to determine the area and perimeter of two-dimensional figures, such as parallelograms, trapezoids, and circles. In addition, students should be able to determine the volume and surface area of prisms and cylinders.

The topics covered in these problems were chosen from state and national standards:

- Understand the proportional relationships between units of measurement (same-system conversions)
- Apply formulas for determining area and perimeter or circumference of parallelograms, trapezoids, and circles
- Determine the surface area and volume of rectangular prisms and cylinders

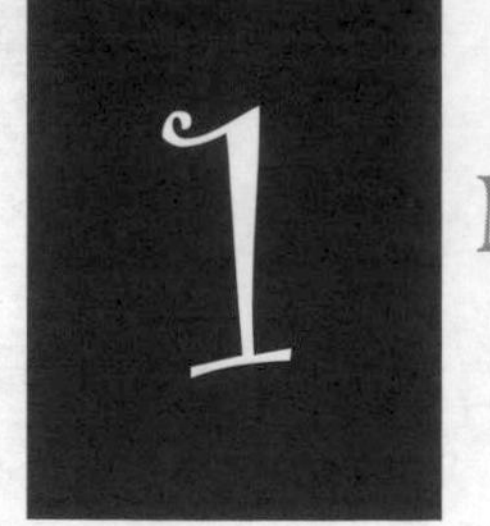

Measurement

Problem

Puffy the Hamster Runs for His Life

Sadly, the astoundingly rich Snedley Measlebop died without the hand of a friend to hold. His closest friend was his beloved Puffy, a copper-colored hamster. It was Snedley Measlebop's wish, upon his passing, that Puffy's exceptional standard of living continue under the care of Jason, his personal trainer.

Naturally, Puffy was saddened by the loss of his friend. His gold cage was full of sympathy cards that he had not yet eaten and his belly and cheeks were full of the ones that he had. Puffy was spending an unhealthy amount of time burrowing and had lost all interest in his gold running wheel. It was certainly understandable, during this difficult time. But as a result of his inactivity, Puffy became a bit puffier and more than once became wedged in the cardboard toilet paper roll that Jason replaced daily in his cage. Each time he got stuck, Puffy would have to nibble his way out, which added to his sizable problem.

Jason began to worry about his furry little charge. "Puffy," he said softly one day to a mound of wood shavings in the corner of the cage. "It's time to get moving. You're going to run 3.4 meters today."

"N-o-o-o!" came a wail from the slowly moving mass of shavings.

Jason reached in and gently pulled out the fat little hamster. Puffy moaned and struggled for a moment before reluctantly falling into his old routine of doing stretches, sit-ups, and jumping jacks to prepare for his run. He leaned his front paws on the middle bars of the cage and waddled one hind leg back at a time, planting a paw pad for a good life-affirming stretch.

Jason cleared a 3.4-meter track for Puffy's run and set the chubby hamster on the starting line. Puffy rolled onto his butt and began to sob.

"I can't run 3.4 meters," he sniveled as tears dripped from the end of his whiskers. "My paw pads are sore, I can hardly breathe, and I have an entire sympathy card in my left cheek!"

"Then now's the time, Puff," said Jason. "You get up on your feet, you furry little eating machine, and kick it out, one centimeter at a time if you have to."

And Puffy did! Quite slowly at first, but centimeter by centimeter he moved. Jason, red-faced, cheered, yelled, cajoled, jumped, stomped, and ran along beside, waving, singing, praising, hoping, throwing his hands in the air and ranting! Jason moved to the finish line, yelling, "Kick it out, Puffy! You're the man! Well, no you're not, but you're the hamster!" Puffy kept going, faster now, closer and closer.

"Go, you waddling wonder!" Jason shrieked at the top of his lungs. Puffy's paws were on fire! And across the line he came!

He high-fived as best he could and fell onto his back with a smile.

He ran the 3.4 meters. To really make Puffy feel good, can you tell him how many centimeters he ran?

Driving Me Crazy

Toshia, Alley, and Thomas E are wonderful children. I love them like raisin toast and I'm pleased as punch to be their mom. But riding in the car with them can be quite challenging. I believe they have a private agreement to complain every 12,144 ft.

"Can you turn up the CD player? I can't hear the story."

"Can you put the air on?"

"Can I have a blanket?"

"Mo-om, Thomas E is poking me with a pencil."

"Thomas E, stop poking her with a pencil."

"I'm not. I'm pretending it's a sword."

"Oh, that's different."

"Can you turn the air off?"

"It stinks back here."

"Pull over, there's a fly in the car!"

"I need a pencil to do my homework."

"Disarm, Thomas E."

"Can I have the snack bag?"

"Which bag has more chips?"

"It doesn't matter. They're the same."

"Can we stop at a store and by a scale?"

Kids have traveled this way throughout history. Every 12,144 ft., after they yelled "Wagons ho!" pioneer children could be heard yelling, "Can you turn up Grandpa? I can't hear the story."

"I'm hot—can you make the horses run?"

"Mo-om, Charles took off my bonnet and hit me with the iron skillet."

"No I didn't, I'm pretending her head is an egg."

"It stinks back here."

"Which pot has more mush in it?"

"Pull over, there's a fly in the wagon!"

How many miles do they wait before beginning to pester again?

Problem

Selling Tights to a Porcupine

Dave had left his previous employment at the Circus Supplies Store for an opportunity to make his mark in an advertising job. Dave liked to sell things. It was said that he could sell tights to a porcupine. His motto at Abracadabra Advertising was "We sell your product whether it's good for people or not!"

Answering the doorbell one afternoon, Dave found Ivey Burns, who had invented a computer game board that was powered by a stationary bicycle. It's a great little gizmo that gives kids the computer game time that they love with the energy and fitness that comes from bike riding. Ivey had a little bit of money for advertising, so she was meeting with different advertising companies to see which might do a good job.

Dave explained to her that he was very good at advertising and that, although he didn't want to brag, he had created the advertising campaign for Joe's chili, with the slogan "So hot to eat, you think you're gonna be sick and then you're not."

"Did it sell a lot of chili?" asked Ivey.

"I don't know. I bought some. I ate it without getting sick, too," said Dave.

"Very impressive," said Ivey.

"Let me show you the brainchild I've got for you. We're gonna use parallelogram-shaped billboards! Everybody uses rectangular billboards, but ours'll be parallelograms. It'll grab people's attention cause they'll think it's a regular rectangular billboard that's about to fall on top of them! Your gizmo will go like hotcakes!"

"Hmm," said Ivey. "Show me what that would look like."

Dave drew a parallelogram like this.

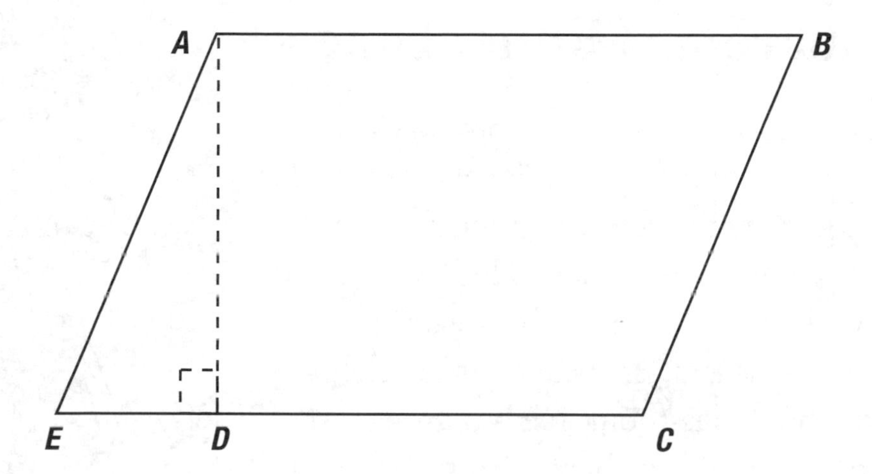

"What's the area of that figure?" asked Ivey, intrigued.

Uh-oh, Dave doesn't know how to find the area of a parallelogram. Write what lengths he needs to know to figure out the area. Quick, before he tells Ivey about his dog-food campaign slogan, "Most dogs eat trash. Surely they'll eat this stuff!" and loses her altogether.

Problem

Nightmare at Circle Time

Leonardo Digit's success on the California Angles geometry team made him a bit of a local celebrity. He felt a little awkward with all of the attention, but he also enjoyed it. He deserved it, too. He never missed a practice, gave 110%, and played his heart out.

Today Leonardo is a guest speaker at Teacher Elizabeth's preschool classroom. He has brought a rather beautiful trapezoid to share with the students, who are all seated in a circle on the rug before him. He has regaled the children with information about the trapezoid and now they are asking questions.

They shoot their little hands up into the air, hoping to be called on.

"Yes," says Leonardo with a smile, pointing to a little yellow-haired girl with her sandals on the wrong feet.

"Um... um... um... I forget," she says, twisting her shirt up over her head.

Leonardo tells her that, when she remembers, he'll call on her again. "OK," comes her muffled reply.

Then Leonardo calls on an intelligent-looking boy who still has grape jelly on his face from breakfast.

"How big is your trapezoid?" asks the jelly-faced boy.

"Good question," says Leonardo. "The bases are 6 inches and 8 inches, and its height is 5 inches, so the area is..."

"My dad is inches tall," interrupts a boy who has been wiggling since Leonardo began.

"How many inches tall is he?" asks Leonardo.

"What do you mean?" replies the boy.

"Well," explains Leonardo as patiently as he can, "You don't just say 'inches' tall. There's a number in front of 'inches' that tells you how many."

"Oh," said the wiggly boy, "My dad doesn't have a number. He just has inches. He's bald."

"My dad is more bigger than that," shouts a girl with butterflies on her headband.

"Thank you for sharing," says Leonardo, quickly trying to reestablish control. "Now, back to my trapezoid, the area is . . ."

"Ooh. Ooh. Ooh!" exclaims a boy with unusually large ears and a missing tooth, waving his hand a fraction of an inch from Leonardo's nose.

"Yes, someone with their hand up," Leonardo says, reluctantly pointing at the boy.

"How come you have that pointy bump on your neck that goes up and down?" the boy asks.

"That's my Adam's apple," answers Leonardo.

"Eeeeww!" choruses the whole class.

"You should chew your food and swallow it all of the way," yells Jelly Face.

"How can it be yours if it's Adam's?" asks an adorable little girl with her finger up her nose.

"You're a stealer," shouts another.

"I'm stuck in my shirt," wails the yellow-haired girl.

"When is it recess?" demands one of them.

"Ouch! The chair leg is in my eye," whines one.

"I'm telling Teacher Elizabeth you took Adam's apple," threatens someone.

Now Leonardo is sweating. This is harder than big league geometry. He remembers he saw a movie once where birds joined in, a few at a time, to peck a guy to death. This is a lot like that.

You're his only hope. Tell kids the area of the trapezoid and get him out of there.

5 Measurement

Problem

Boom Goes the Great Zamboni

In the aftermath of one of the worst circus disasters in the history of the art form, the Great Zamboni began to seek recovery.

"Circus Supplies—Everything essential for entertaining exquisitely under the big top, conveniently located in Pontiac, Michigan. Thiz Dave speaking. How may I enhance your circus?" the Great Zamboni heard through the phone receiver that he could barely hold to his ear as he lay bandaged like a mummy in his bed.

"This is the Great Zamboni," he answered with great effort. "I am calling to seek satisfaction from the Circus Supplies Store for the collapse of your faulty cylinder on which I, the Great Zamboni, had stood as ringmaster of the Great Zamboni's World-Famous Circus. The collapse of your cylinder, sir, caused my own collapse, which startled Bongo, the driver of our clown car, which caused him to crash, injuring thirty clowns and resulting in thirty no-seat-belt traffic tickets. The clown car crashed into Madam Tutuni's Death-Defying Dance of the Big Cats Show, startling one of the big cats, who then refused to leap through the big hoop and instead sprang at me, practically tearing me limb from limb."

"You were attacked by a tiger?" marveled Dave.

"No, it was not a tiger," the Great Zamboni muttered.

"A lion?" Dave asked, amazed.

"No. If you must know, it was a tabby, but it was puffed up quite large and seemed to think the Great Zamboni was a fish," sniffed the Great Zamboni. "Now, you must refund my money for your faulty cylinder immediately!"

"People don't really like the animal circuses anymore anyway," said Dave. "We may have done you a favor. You know, the Chinese Circus does a wonderful show with a unicycle. They catapult one acrobat after another on top of the shoulders of Ming Chow, the unicycle rider. It's fantastic. It keeps costs down, too, because they only have to purchase and maintain one unicycle.

"Maybe you could get rid of the animals and your animal trainers could learn tricks. Your elephant keeper could learn a unicycle act. Honestly, he'd probably like it better than shoveling..."

"My elephant keeper is 65 years old and weighs 300 pounds," Zamboni interrupted in a sputtering rage.

"That'd make your circus unique," persisted Dave the salesman. "You'd be the only circus with an out-of-shape older gentleman on a unicycle. I don't think it would bring audiences from miles around, but people in the immediate area with nothing to do might drop in.

"Now, for a minimum down payment, I could set aside a unicycle for you today."

"I don't want a unicycle. I want my money back," demanded the Great Zamboni.

"I tell you what," Dave chirped happily into the phone. "I'll make you a straight trade—the cylinder for an extrawide trapeze."

"No, no, no!" screamed the Great Zamboni, his stitches starting to pop.

"Are you sure you bought it here? What does it look like? Tell me the surface area and the volume and I'll see what I can do.... Mr. Zamboni? Are you there?"

You'd better take over for the Great Zamboni. The radius of the base of the faulty cylinder is 5 feet and the height of the cylinder is 10 feet.

A. Find its surface area.

B. Find its volume.

6 Measurement

Problem

Pssst—It's Me, Pssst

Agent Pssst from Rhubarb stood peering over his open newspaper outside of the top-secret government headquarters in Rutabaga, just like it said to do in his spy manual. Because of the unfriendly relations between the small country of Rutabaga, where he now stood, and Rhubarb, where he lived, both countries had dispatched agents to spy on each other. Rhubarb produced milk, and Rutabaga produced cookies, but now neither country would agree to export its product to the other nation. Pssst's mission was to watch for anything suspicious going on, and he was pretty sure he had seen something worth reporting. He raised his wrist to his mouth slowly, trying to appear as though he was just wiping cookie crumbs off his mouth.

"Pssst to Headquarters," he whispered into his watch.

"Headquarters here. Who's this?"

"Pssst," said Pssst.

"Yes, we're here. You've reached Headquarters. Who's this?"

"Pssst," answered Pssst.

"What?" his wristwatch replied.

"It's Agent *Pssst*!" Pssst shouted into his watch.

Heads appeared in every window of the top-secret government building, and a parade of armed guards trooped out of the door and headed straight for Agent Pssst, whose cover had been blown. Pssst dropped the paper and ran, just like it said to do in his spy manual.

He dashed across the street into a women's shoe store and pretended to shop for shoes, just like it said to do in his spy manual.

"May I help you?" the store clerk asked, eyeing him suspiciously.

Pssst couldn't leave the store without reporting the very fishy rectangular prism with a volume of 300 cubic feet he had seen carried into the top-secret government building by a bunch of burly Rutabagans. Thinking quickly, he stammered, "I was looking for some shoes . . . for uuuh, my wife. Could I try these on? We have the same shoe size." And he randomly grabbed a fancy open-toed shoe with a three-inch heel. Luckily, the store had the shoe in a size 13 double wide. Pssst tried them on and stumbled through the store, leaning from shelf to shelf.

"They're perfect!" Pssst pretended, stalling for time. "Do you have them in pink?"

"The walker is sold separately," the clerk mumbled on his way into the back.

Pssst quickly made his report, and Headquarters wanted to know the dimensions of the rectangular prism. He had no idea what they were.

"Dimensions? I don't know the dimensions! I've almost snapped both of my ankles just to get you the volume! There's nothing in the spy manual about dimensions! It specifically says, 'When reporting the observing of a fishy rectangular prism, note volume'!"

Could you help him out? Find the dimensions of a rectangular prism with a volume of 300 cubic feet. In case this happens again another time, can you find a different prism with this volume?

Extra! Extra!

Measurement is important. Without it, people wouldn't know their shoe size. Here are some extra problems to help you avoid blisters.

1. **How many kilometers are there in 56.3 meters?**

2. **How many inches are there in 83 yards?**

3. **A parallelogram has an area of 10 square inches. Using a ruler, draw such a parallelogram. Is there more than one possibility?**

4. Find the area of a trapezoid with a height of 8 feet and bases of 9 and 12 feet.

5. Find the surface area and volume of a rectangular prism whose base is 4.5 ft. by 6 ft. and whose height is 15 ft.

6. Two cylinders have the same heights, but one has a radius that is twice the other. How will their volumes compare?

Measurement

Teacher Notes

1. Puffy the Hamster Runs for His Life

To answer problem 1, students need to know that there are 100 centimeters in a meter. The next task is to decide whether to multiply 3.4 by 100 or divide 3.4 by 100. It's important for students to think about whether the answer should be less than or greater than 3.4. The number of centimeters should be greater than the number of meters, since a centimeter is smaller than a meter, and therefore it makes sense to multiply 3.4 by 100. The answer is 340 centimeters.

Students should feel comfortable multiplying and dividing by powers of 10 without using a calculator or pencil and paper. Another way to think about this problem is to set up a proportion that reflects the conversion factor:

$$\frac{100 \text{ cm}}{1 \text{ m}} = \frac{x \text{ cm}}{3.4 \text{ m}}$$

Multiplying both sides of the equation by 3.4 m gives the answer of

$$\frac{100 \text{ cm}(3.4 \text{ m})}{1 \text{ m}} = 340 \text{ cm}$$

Students will encounter these kinds of conversions throughout their careers, in both science and mathematics.

2. Driving Me Crazy

Problem 2 requires the same kind of thinking as problem 1 in order to convert from feet to miles. There are 5,280 feet in a mile. Again it's important to estimate whether the answer should be less than or greater than 12,144. Since feet are smaller than miles, it makes sense that the answer will be smaller and therefore to divide 12,144 by 5,280 to get the number of miles. The answer is

$$\frac{12{,}144 \text{ ft.}}{5{,}280 \text{ ft./mi}} = 2.3 \text{ miles}$$

We can also set up a proportion that reflects the conversion factor

$$\frac{1 \text{ mi.}}{5{,}280 \text{ ft.}} = \frac{x \text{ mi.}}{12{,}144 \text{ ft.}}$$

Multiplying both sides of the equation by 12,144 ft., we get

$$\frac{(1\text{mi.})(12{,}144 \text{ ft.})}{5{,}280 \text{ ft.}} = 2.3 \text{ mi.}$$

Note that the units are also divided to obtain miles in the answer.

3. Selling Tights to a Porcupine

The simplest way to think about problem 3 is to realize that the area of a parallelogram is found by multiplying the base by the height. In the drawing, a height (*AD*) has been drawn to base *EC*. Therefore if we knew *AD* and *EC*, we could find the area of the parallelogram.

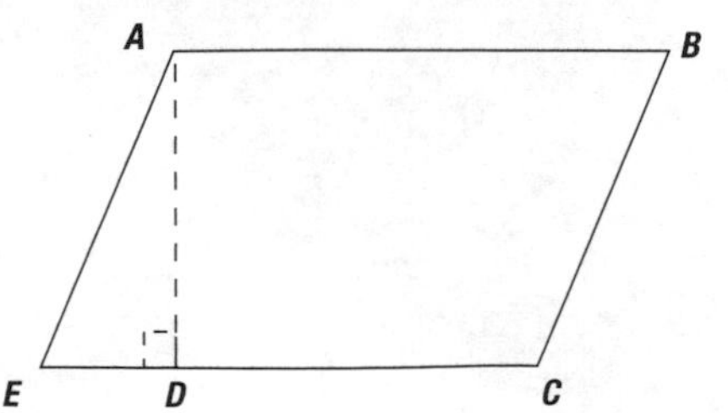

As a challenge, ask students if there is a way to find the area if they know the lengths of AE, ED, and AB.

4. Nightmare at Circle Time

For problem 4, students need to know the formula for the area of a trapezoid: $\frac{1}{2}h(b_1 + b_2)$, or one-half the height times the sum of the bases. Therefore, the area would be $\frac{1}{2}(5)(6 + 8)$, or 35 square inches.

An interesting way to think about this area formula is to consider a dissection of the trapezoid. That is, cut the trapezoid to form another polygon so that area is preserved. Any trapezoid can be dissected into a parallelogram.

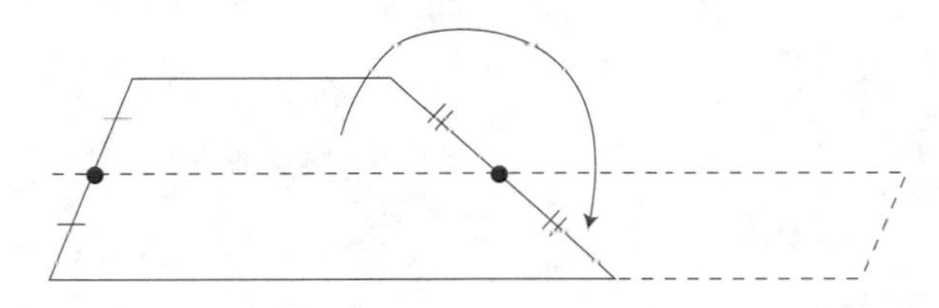

Find the midpoints of the two nonparallel sides, connect them with a line segment, and cut along the line segment. Now rotate the upper half clockwise so that the two congruent sides coincide. A parallelogram is formed, with one set of parallel sides equal to the sum of the bases of the original trapezoid. The height of the parallelogram is half the height of the original trapezoid. Therefore to find the area of the parallelogram (base times height), we are multiplying the sum of the original trapezoid's bases and half of the original trapezoid's height. As an extension, ask students if they can dissect any triangle to form a parallelogram.

5. Boom Goes the Great Zamboni

For part A of problem 5, students need to determine the surface area of a cylinder. If they do not remember the formula, ask them to think about taking a sheet of paper and folding it to form the lateral surface of a cylinder. In fact, the lateral area is the area of a rectangle, where one of the dimensions is the same as the height of the cylinder, and the other is the circumference of the circle that forms its base.

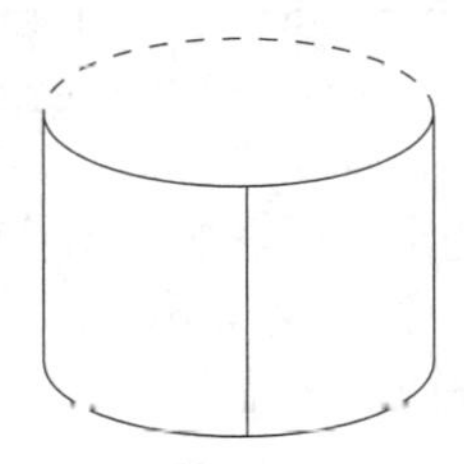

Since the radius of the base is 5 feet, the circumference is $2\pi r$, or 10π. To find the lateral area, then, we multiply 10π by the height, 10, to get 100π. We now need to add in the area of the two bases, which are circles, each with area πr^2, or 25π. Added together, their areas equal 50π. Therefore the total surface area is $100\pi + 50\pi$, or 150π square feet. If students use 3.14 as an approximation for π, the answer would be 471 sq. ft.

For part B, we need to find the volume of the cylinder. It makes life much easier to find the volumes of prisms in general if students understand that the volume of a cylinder, like the volume of any other prism, is the area of the base times the height. Since the base of a cylinder is a circle, we need simply to find the area of the circle (25π) and multiply by the height (10), to arrive at the answer of 250π cubic feet for the volume or 785 cubic feet if 3.14 is used for π. As an extension, ask students to find a rectangular prism with the same volume as this cylinder.

6. Pssst—It's Me, Pssst

To answer problem 6, students must work backward to find the dimensions of a rectangular prism given its volume. The kind of thinking required to do this type of problem is different than simply applying a formula for volume. First, students need to know where the 300 cubic feet came from—it's the product of the area of the prism's base and its height. Therefore, they need to find two numbers that multiply to 300. Obviously, there are many solutions here, even if we restrict ourselves to whole numbers. For example, the area of the base could be 30 square feet and the width could be 10 feet. If the area of the base is 30 square feet, the length could be 6 feet and the width could be 5 feet. Therefore the dimensions could be 5 ft. × 6 ft. × 10 ft. As for other prisms with the same volume, another could be 4 ft. × 5 ft. × 15 ft.

Students may find it interesting to think of how this problem relates to the prime factors of 300: 2 × 2 × 3 × 5 × 5 (we would want to include 1 and 300 as well, since one possible prism has dimensions of 1 × 1 × 300). As an extension, ask students to determine how many possible prisms there are with a volume of 300 cubic feet if the dimensions are whole numbers.

Extra! Extra!

1. .0563 kilometers
2. 2,988 inches
3. 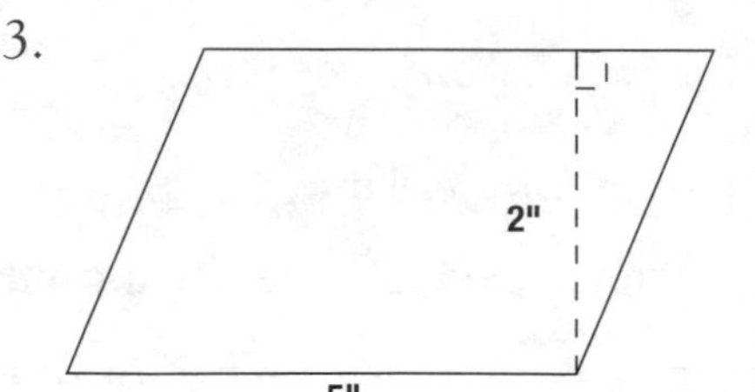

   Yes, there are an infinite number of possibilities.
4. 84 square feet
5. surface area: 369 square feet; volume: 405 cubic feet
6. The volume of the larger cylinder will be four times that of the smaller cylinder.

Data Analysis, Statistics & Probability

The problems that follow are in the Data Analysis, Statistics, and Probability strand. The mathematics in these problems focuses on developing students' data analysis skills as they describe data sets using measures of central tendency. Students in grades 6 and 7 learn to construct and analyze different representations of data, such as stem-and-leaf plots. This strand also includes the study of probability as students determine outcomes and probabilities for simple compound events.

The topics covered in these problems were chosen from state and national standards:

- Describe data sets using median, mean, mode, and range
- Construct stem-and-leaf plots, line plots, circle graphs, and Venn diagrams
- Predict the probability of outcomes of simple compound events, such as multiple coin tosses

Big Stirs

The Exaggerators Club meets Wednesday afternoons in the small town of Rupertsville, or "practically every day in the smallest town in the world," according to their poster in the library. Although the club includes some topflight exaggerators, their math skills are a bit limited.

They spent a fun but grueling two hours today trying to impress a panel of judges with their stirring skills, hoping that the swirl size, dissolving, beverage containment, presentation, and degree of difficulty of their stirs would get them selected for their local amateur competitive stirring team. With the tryouts over, they sat in the lobby of the gymnasium, anxiously awaiting their scores.

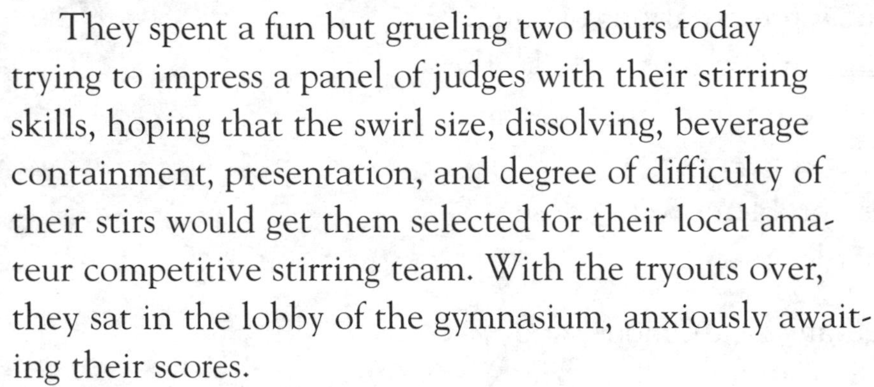

"Sorry fellas, but yours truly may have to drop out of the Exaggerators Club. I just don't see how I can be on the road with the red-hot Rupertsville Stirring Team, work out 20 hours a day, read 3 nonfiction books each day, work a 50-hour workweek, and meet with you slackers once a week, no matter how much good it does you. After all, my SUV only goes up to 200 mph," bragged Tell-It-Tall Totter.

Rosetta Donut waved him off. "You're a great guy, Tell-It-Tall, and I wish I could help you more, but after my 100-point stirring performance, I'm gonna be busy doing spoon commercials."

Wanda Whatnot was covered in milk. "How did I do it, you ask?" she blurted out, although no one had asked. "I use my hips! I stick out my right hip and stir twice to my left and then I stick out my left hip and stir twice to my right—and-a-boom-and-a-boom-and-a-boom-boom-boom—and I got 200 points easy," she said, swinging her hips and waving her imaginary spoon wildly in the air with her eyes closed.

"And I'm sure you'll get a nice little runner-up prize for trying so hard," Curly Hushup put in, blowing on the fingers of his still revolving hand. "But 300 points puts me on top, according to my finger counting."

Just as the others were puckering up to inflate their own figures, an official from the State Competitive Stirring Association came in with their scores.

"I want to thank you all for your interest in the exciting sport of competitive stirring. You did very well. The scores of the 7 members of your group had a range of 10, a median of 30, a mean of 33, and a mode of 38."

"Oh, I see," said Tell-It-Tall quietly, while sinking into this chair. "I'm used to the way they score in Italy."

Create a possible set of scores for the exaggerating competitive stirrers to fit the official's numbers.

Problem

The Sweet Smell of Success

Marcus was an unbelievably nice guy and a great lacrosse player, but when it came to selling candy bars to raise money for his lacrosse team's supplies and expenses (ice packs mostly), he was not the golden boy. He tried selling to his neighbors. At the first house he forgot to bring up the subject of lacrosse or candy, but the neighbor sold him two tickets to her child's school production of *Annie*, and right there in their living room he got to see the little cherub sing "It's the Hard-Knock Life" four times. It really made him appreciate how good he had it—usually.

The people in the second house Marcus went to bought 2 boxes of candy bars from him, which was great because he needed to borrow that money for a down payment on the 3 boxes of Girl Scout cookies he bought at the third house. Things went much better after that. Once he bandaged the dog bite he'd gotten at the fourth house, he was practically good as new, and the dog's owners even bought some candy from him after he mowed their lawn and carried some boxes up to the attic for them.

Meanwhile, Marcus' teammate Colin sold 78 candy bars during a traffic jam, and people just came up to David on the street to ask if he had any candy

they could buy. He sold 97 boxes. Andy's mother sold 82 boxes to the cheaters at her weight-loss clinic, but Marcus just wasn't like that. When his neighbor with an unhealthy weight wanted a box, Marcus bought it and ate if for him so as not to add to the guy's problems. Then when Marcus went to visit his great-uncle Dick at the retirement home, the entire community of 50 senior citizens went in on one box.

The table below shows the number of candy bars sold by the boys on Marcus' lacrosse team.

BOY	NUMBER OF CANDY BARS SOLD
Marcus	11
Ian	79
Anson	27
Dan	21
Matt	91
Gray	18
Colin	78
Scott	15
David	97
Andy	82
Ben	24
Chris	31
George	20
Greg	68

A. Which would be better, a line plot or a stem-and-leaf plot, for displaying the values for the number of candy bars sold? Explain your reasoning.

B. Create the plot you chose in part A.

C. Find the range and median number of candy bars sold.

Mutiny on *The Booty Bucket*

Thirteen-year-old Armando stood quaking at the end of the plank with sharks snapping in the churning sea beneath him and the Jolly Roger snapping in the wind on the flagpole above. Captain Stubbly Beard stood on the deck of his weatherworn ship, *The Booty Bucket*, waving his sword to indicate that the sharp end of it was Armando's alternative, should he not plunge to his death off the plank. It was a tough choice. As Armando paused to quake a bit longer, the noisy discontent of the crew could be heard growing behind him.

"We'll stand no more of it, Stubbly Beard!"

"I say let's keelhaul him!"

"Let's feed his fingers to the creatures of the deep."

"Let's be very, very mean to him!" The shouts came louder and louder from one after another gravelly voice among the filthy cutthroat band of sea robbers seething on the deck behind him.

"Where's the treasure?" yelled a guy with a mouth full of green teeth.

"We want to pillage!" shouted another guy with no teeth.

"You promised us fresh eye patches!" hollered yet another, who in fact wore a heavily soiled eye patch.

Armando thought the situation had all of the earmarks of a mutiny. Here he saw a slight glimmer of a chance to save himself.

"Captain Stubbly Beard," he called, straining his voice to be heard over the rabble. "It sounds as though your crew is dissatisfied."

"It would seem so, my young shark appetizer. What of it?"

"They're giving you a lot of information about the reasons for their discontent. I can help you organize and analyze your data, and perhaps you can avoid being hung or thrown overboard, neither of which seems pleasant," Armando shouted, inching toward the relative safety of the deck.

"Oh, ye can, can ye?" replied Stubbly Beard suspiciously.

With Stubbly Beard's grudging permission, Armando marched forward and questioned the men one at a time about their grievances. A lot of them wished to have more free time. Some wanted greater access to the rum supply. Some hoped to learn more pirate songs. Billy the Bilge Rat wanted to choose new nicknames. Lots of them wanted more treasure. Some were just plain sick of seafood. One guy wanted soap.

Breathing through his mouth all the while to avoid the stench of the angry pirates, Armando carefully recorded all of their responses and organized them on a line plot for better understanding.

He really wanted to find the response that came up most often. How could he find the mode of the set of data from the line plot?

Numbers of Numbers for Sale

The numbers practically flew off the shelves and into the customers' hands at Dick Digit's Number Depot ("Where the whole family can shop for numbers and operations don't hurt a bit!"). Leonardo Digit frantically staffed the cash register, restocked shelves, and helped customers find the numbers that were right for them, to help his dad, who was *the* Dick Digit. It's a good thing he knew how to juggle. The place was swamped today and who knows why? Maybe it was because of Groundhog Day. Maybe it was because of the new attention-grabbing 4 they had hung on the front of the building, or maybe the excitement was spreading about the time-saving convenience and beauty of the self-cleaning 100 they had recently begun to sell, but Leonardo had never seen the store so busy. The cash register even broke, which Leonardo's dad said was "probably a good thing if you look at it the right way." Some stores would have had to close for the day if their cash register broke. But fortunately, Leonardo had grown up around numbers. He had an easy, friendly relationship with them and could calculate almost as fast as the machine. Still, it was not always where his heart was.

After the last customer was finally out the door, Leonardo grabbed his jacket to leave and practically tripped over his dad, who was in the middle of an aisle, slumped inside a big rubber 0, looking quite miserable.

"Son, I think I hurt my back while getting a fifty-pound 62 off a high shelf. I'm going to need you to collect and organize some data so that I can place some orders to restock the store."

"Oohh," moaned Leonardo, reaching up to hang off a bright pink number 2 that was suspended from the ceiling. "Dad, did you know that Emmett Kelly, the famous clown, started out as a trapeze artist?"

"What a coincidence," said his dad, pretending to be excited while he sunk farther into the rubber 0. "And you're starting out helping your dad collect and organize data about number sales! It's practically the same thing, if you look at it the right way."

Leonardo couldn't help wondering if his dad looked at things upside down and blindfolded, but he dutifully made a list of the items sold and then organized the data in a stem-and-leaf plot while balancing a beautiful ceramic 1 on his head.

How did he find the mode from his stem-and-leaf plot? Describe a general process for finding the mode from a stem-and-leaf plot.

Venn Can We Be Friends?

The tensions between the tiny countries of Rhubarb and Rutabaga had reached a feverish pitch. Rhubarb no longer exported its milk to Rutabaga and Rutabaga wouldn't be caught dead selling its cookies to Rhubarb. Their armies were toe-to-toe at the border and both sides had infiltrated the other with spies. The truth is, Rhubarbite spies had discovered that not a whole heck of a lot was going on in Rutabaga and Rutabagan spies had uncovered a similar lack of activity in Rhubarb, but neither country could really admit it. There is a history to their conflict that includes unpardonable insults on the part of both sides, but it belongs in dusty old history books, not here.

At this point, there was a very real possibility that war might break out between the two nations. They were each like a bomb that could detonate over the slightest piffle. Fortunately, the Worldwide Peaceful Council of Math Doers selected you to meet with the leaders of the two countries to bring a peaceful end to the hostilities between them and their people. Although they each entered the room

backward to avoid looking at their enemy, within seconds, there was one on each side of you, pounding on the table and screaming at the other.

The Rhubarbite leader wore a loose-fitting white robe with a cowbell around his neck, which he occasionally rang over your head. The Rutabagan leader wore a stiff red suit with gold piping around the seams and a stiffly waxed mustache that was so long and thin he often walked sideways to avoid bumping it on things. He wasn't just yelling; he was frothing, and bits of his spittle were flying about your head.

You had had enough. You jumped up on the long table, took two running steps, and slid down to the end. You grabbed a dry-erase marker, and without taking your eyes off the battling leaders, you drew two circles on the whiteboard, or so you thought. When you noticed a strange look on the two men's faces, you looked at the board and saw that it was empty. You grabbed another marker, made sure it wasn't dried out, and quickly drew two intersecting circles. You labeled one of the circles "Both Pairs of Opposite Sides Are Parallel" and the other "All Angles Are Right Angles." Of course, you were creating a Venn diagram. You asked the two men, who were by now looking at you quite oddly, where in the circles each of the following should be placed:

parallelogram
rhombus
rectangle
square
trapezoid

When the Venn diagram was complete, you began nervously, "Gentlemen, we are all a part of a Venn diagram. Some characteristics we may share with our neighbor, our countrymen, or those in a foreign land beyond our borders, and some we may not. But there's a bigger imaginary circle around these two that could be labeled 'Shapes,' and they would all belong inside. And there's a bigger circle that surrounds both of your countries as well as the rest of the world, and it could be labeled 'People,' and we all belong."

For a moment there was silence. The Rhubarbite leader removed his cowbell and quietly handed it to the Rutabagan leader, who gently clipped the ends of his pointy waxed mustache and pressed them into the outstretched hand of the Rhubarbite leader, with whom he shared the circles of neighbor, coworker, fellow human being, and friend.

Great job! By the way, could you re-create the Venn diagram of shapes?

A Brilliant Stroke of Dumb Luck

This was a bad year for International Competitive Stirring. It's a glorious sport of skill, speed, and flourish that normally brings such joy to the players and fans alike. This year somehow it hadn't. Maybe it was the stress of global warming and the growing concern about the shrinking polar ice cap, but people were edgy.

The finest stirrers from countries all over the world marched into the arena with grandeur, spectacle, and pride. The judges sat poised to evaluate each athlete's stirring performance for speed, swirl size, dissolve, beverage containment, and degree of difficulty. Following the official protocol for stirring competition, an official came to the center of the arena, among hundreds of competitors and thousands of spectators, and tossed a coin to fairly determine which team would go first. It was heads, but before the official could finish announcing that the team from Jaga would go first, a dark cloud of discontent seemed to come over the room. Athletes, coaches, fans, and cotton candy vendors began shouting that the decision was unfair.

"We want to go first!" shouted a coach from a country whose name I am embarrassed to mention (but I'll give you a hint: it has the initials USA).

"We usually get heads!" whined a stirrer from the Bulltonian team.

One of the cotton candy vendors just went off. No one really understood what he was saying, but he got sticky stuff all over himself in the process. In no time at all, pretty much everyone was shouting at one another or at least forgetting to use the magic words. Well, not everyone. Bartholemew the Baby Genius' mother sat aghast at what she was seeing and hearing. Feeling strongly that this sort of scene was not what she wanted her innocent son to witness, she turned to remove him from his booster chair in the seat beside her—and he was gone! Her panicked cries couldn't even be heard over the sound of the angry crowd, but in a moment, to her amazement, she heard her baby's familiar gurgling over the public address system! Bartholemew the Baby Genius had crawled all of the way to the sound booth and was now blowing loud raspberries into the microphone, which brought a curious hush to the crowd.

"I have something that I think is important to say," he said," and I'd like you to listen because I'd like to get my pacifier back into my mouth as soon as possible."

Bartholemew the Baby Genius then told the puzzled crowd what the probability is of a coin that is tossed twice coming up heads both times, what the probability is of it coming up tails both times, what the probability is of it coming up heads the first time and tails the second time, and what the probability is of it coming up heads at least once.

By now people had begun to point with amazement toward the glass window of the sound booth where Bartholemew sat.

"It's good to know probability, but the point is, you can't control it. Things can go against you even when the odds are in your favor. What you *can* control is what you do when you get tails and you wanted heads. Do you stir for all you're worth, maybe even harder to make up for the disadvantage of not starting first? Or, better yet, do you stir just for the sheer joy of it? Or do you let this thing that you can't control decide what kind of stirrer you're going to be?" With that, Bartholemew popped his pacifier back into his mouth and, with a quick, satisfying suck, crawled back to his mother.

People in the crowd began nodding appreciatively. In a moment, the officially selected starting stirrer, chosen with the random toss of a coin, ran out to his waiting tall glass of milk and carefully measured the powder mixture to begin the preliminary rounds of next year's International Stirring Competition.

"Let the thrills and spills of stirring begin!" shouted the announcer over the clamor of a jubilant crowd.

What were the probabilities Bartholemew told the crowd?
When a coin is tossed twice,

A. What is the probability it will land heads both times?

B. What is the probability it will land tails both times?

C. What is the probability it will land heads the first time and tails the second time?

D. What is the probability of getting at least one head?

Extra! Extra!

What is the probability of getting you to do some extra Data Analysis, Statistics & Probability problems? You guessed it.

1. **Find a set of 5 values that fits the following description: The range is 20, the median is 15, the mean is 19, and the mode is 12.**

2. The following table shows the number of raffle tickets sold by students on a math team:

NAME	NUMBER OF TICKETS SOLD
Noah	150
Sam	149
Helena	135
Matt	139
David	178
Maya	190
Lydia	104
Nick	157
Zeke	143
Katie	148
Lucy	188
Sarah	149
Zoe	106
Gary	133

A. Create a stem-and-leaf plot for the data.

B. Find the range and median number of candy bars sold.

3. Create a two-circle Venn diagram. Label one of the circles "Quadrilateral" and label the other circle "Not a Triangle." Place each of the following in your Venn diagram:

parallelogram hexagon rhombus trapezoid triangle pentagon

4. One die is tossed twice.

A. What is the probability of getting a 2 each time?

B. What is the probability of getting at least one 2?

C. What is the probability of getting no 2s?

Teacher Notes

1. Big Stirs

For problem 1, students may use a variety of approaches in thinking about how to create the data set. Since the range is 10, the difference between the least value and greatest value must be 10. If there are 7 values, and the mean is 33, the total of the 7 values must be 7×33, or 231. (The mean of a set of values is the sum of those values divided by the number of values.)

Trial and error will probably be the method most students use to find the seven values. If we choose 28 as our least value and 38 as our greatest value, then we know the sequence must be

28 ___ ___ 30 ___ ___ 38

If the greatest value is 38, and we know the mode is 38, that means there must be at least two values of 38. (The mode of a set of values is the one that occurs most often.) Our sequence is now

28 ___ ___ 30 ___ 38 38

The mean is 33, so we currently have $231 - (28 + 30 + 38 + 38)$, or $231 - 134$, or 97. Therefore we need three more values that add to 97, two of which are between 28 and 30 inclusive, and one of which is between 30 and 38 inclusive. One possibility is 29, 30, and 38, forming the set 28, 29, 30, 30, 38, 38, 38. There are certainly other answers (an infinite number of them!). Here's a challenge question for your students: Can you add an eighth value to the set so that the same conditions (for range, mean, and mode) hold?

2. The Sweet Smell of Success

In problem 2, students are given a table of data that includes a number of different values for candy bars sold. Part A asks students to determine the best way to display that data. In general, line plots are useful when values are repeated and the range of values is relatively narrow. In this situation, since the data set contains many different values that are spread out, a stem-and-leaf plot would be more helpful.

The stem-and-leaf plot for part B would be:

Stem	Leaves
0	
1	1 5 8
2	0 1 4 7
3	1
4	
5	
6	8
7	8 9
8	2
9	1 7

In answer to part C, the range is the difference between the least value and the greatest value: $97 - 11$, or 86. The median number of candy bars sold is the middle value when the values are placed in order. However, since there are an even number of values, the median is the average of the two middle values (27 and 31), or 29. As an extension, ask students to add another value that will keep the range the same and the median 29.

3. Mutiny on *The Booty Bucket*

In problem 3, to find the mode of a data set displayed in a line plot, we can look at the tallest stack of marked values. For example, in the following figure, the mode is 17, since that value has the highest stack of *x*'s representing four data points of the value 17.

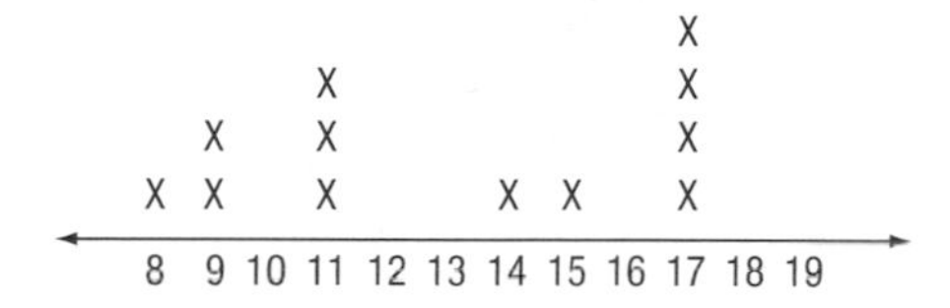

4. Numbers of Numbers for Sale

For problem 4, to find the mode of a data set with a stem-and-leaf plot, we need to look for the digit that appears most often in one stem. In the stem-and-leaf plot created for part B of problem 2, there is no mode since each value appears only once (or some might say each value is a mode!). As students progress through middle school, they will be asked to choose representations that best suit the data given. These kinds of problems are useful in determining the characteristics of different representations.

5. Venn Can We Be Friends?

Problem 5 focuses on Venn diagrams. Venn diagrams were named after John Venn of England in the 19th century. In a Venn diagram with two circles, the circles overlap, as in the following drawing.

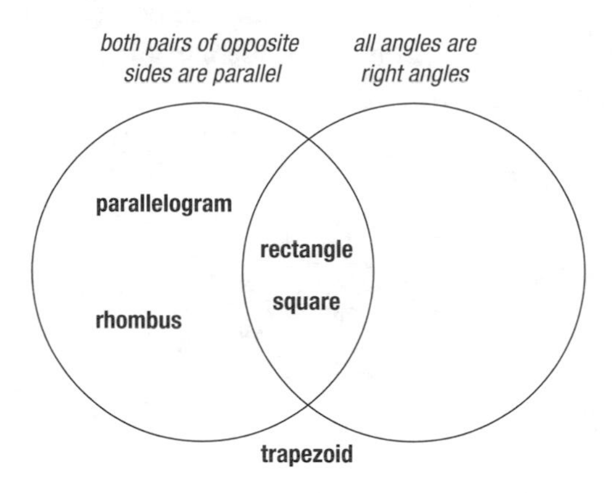

If something is placed inside one of the circles, it must have the attributes of that circle. If something is placed in the overlapping area of the two circles, it must have the attributes of both of those circles. Therefore, *rectangle* and *square* are in the overlapping region, since they each have both pairs of opposite sides parallel *and* all right angles. The parallelogram and rhombus each have both pairs of opposite sides parallel, but they do not have all right angles. Trapezoids have neither characteristic, and therefore they are outside of the circles.

As an extension, ask students to draw a third circle intersecting the other two, labeled "All Sides Are Equal," and place each of these polygons in this new Venn diagram. Here is one more challenge students can play with a partner: Come up with 7 or 8 polygons and categories for a three-circle Venn diagram. Don't tell your partner the categories, but do tell your partner the regions where the polygons lie. See if your partner can guess the categories for the circles.

6. A Brilliant Stroke of Dumb Luck

In part A of problem 6, students are asked to find the probability that when a coin is tossed twice, it will come up heads both times. One way to think about this is to draw all the possibilities for tossing a coin twice. A tree diagram is useful in representing this event:

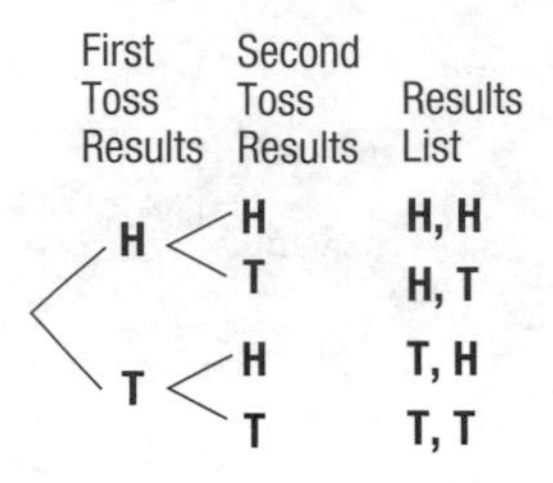

Note that if the coin is tossed twice, there are 4 possible outcomes. Only one of these outcomes is H, H, and therefore the probability of tossing two heads is $\frac{1}{4}$. Another way to show the outcomes of an experiment is to construct an outcome grid:

	HEADS	**TAILS**
HEADS	heads, heads	heads, tails
TAILS	tails, heads	tails, tails

From both the tree diagram and the outcome grid, we can observe that the answer to part B, the probability of getting tails both times, is also $\frac{1}{4}$. The answer to part C, the probability of getting heads the first time and tails the second time, is again $\frac{1}{4}$. For part D, the probability of getting at least one head includes the outcomes of one head or two heads. All of the outcomes have at least one head, except for the outcome of both tails. Therefore the probability of at least one head is $\frac{3}{4}$.

Students will study events with two possible outcomes (binary events) in many different contexts: true/false answers on tests, coin tosses, yes/no responses on surveys, on/off switches, male/female births, and so on. These events have much in common with each other regarding possible outcomes, and analyzing these results is an important aspect of learning probability. There are even connections to Pascal's Triangle! (The first five rows of Pascal's Triangle are

1

1 1

1 2 1

1 3 3 1

1 4 6 4 1

Looking at the third row

1 2 1

the numbers indicate, in order, the outcomes of two heads (1), one head (2), and 0 heads (1), when tossing a coin twice.)

As an extension, ask the same questions as those posed in problem 6, but this time in regard to tossing the coin 3 times.

Extra! Extra!

1. One possible set is 12, 12, 15, 24, 32.
2. A.

10	4 6
11	
12	
13	3 5 9
14	3 8 9 9
15	0 7
16	
17	8
18	8
19	0

B. range: 186; median: 148.5

3.

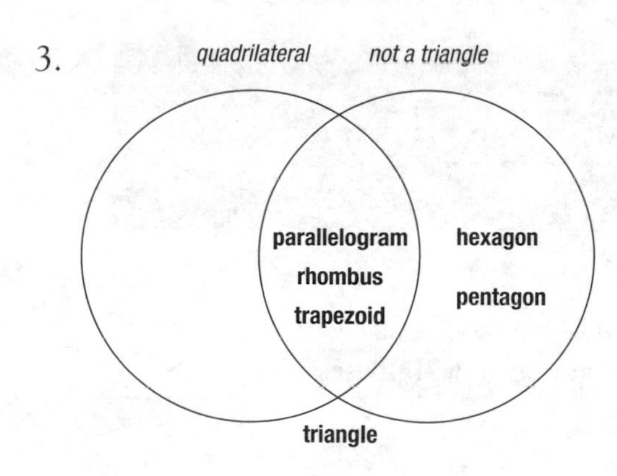

4. A. $\frac{1}{36}$

B. $\frac{11}{36}$

C. $\frac{25}{36}$